THE RENTOKIL LIBRARY

THE WOODWORM PROBLEM

NORMAN E. HICKIN

Scientific Director
Rentokil Limited

HUTCHINSON OF LONDON

HUTCHINSON & CO (*Publishers*) LTD
3 Fitzroy Square, London W1

London Melbourne Sydney Auckland
Wellington Johannesburg Cape Town
and agencies throughout the world

First published 1963
Second edition 1972

Photographs and drawings © Rentokil Laboratories Ltd., 1972,
unless otherwise acknowledged

Produced by Hutchinson Benham Ltd.

*This book has been set in Times type, printed in Great Britain
on coated paper by Anchor Press, Tiptree, Essex, and
bound by Robert Hartnoll Ltd., Bodmin*

ISBN 0 09 113910 4

CONTENTS

PREFACE

It is now eight years since publication of the little book *Woodworm—its Biology and Extermination*, and as the edition is now out of print and a steady demand still remains for it, it has been decided that a new and up-to-date account of the problem should be written. Indeed, the problem today is even more important than in the early 1950's and now constitutes a national problem.

Although much of the earlier material and some of the illustrations remain, there is a large amount of new matter now included. Indeed, during the last eight years there has developed a new technology and industry—the extermination of woodworm from building timbers and joinery, which was only in its infancy when the old book was published, so that some space is devoted in the present book to explain the way this work is organized and conducted. Some of the illustrations used in the reference book on wood-boring insects—*The Insect Factor in Wood Decay*—are included in this present work.

Some notes on insurance against woodworm and on the logical conclusion with regard to new buildings—preservative treatment of carcassing and joinery timber—before erection are given.

Because so much new material is included and the remainder largely rewritten it was thought that it should have a new title—*The Woodworm Problem*.

PREFACE TO SECOND EDITION

A new printing of this book has become necessary, and this has provided the opportunity to make some corrections and amendments to the text in the light of present day knowledge. In this regard I am indebted to my colleague Dr. Colin Hawkes for his valuable comments and suggestions. The colour plate and jacket, and several new black and white photographs, are the work of Mr. Robin Edwards.

N.E.H.
20.2.1972

ACKNOWLEDGEMENTS

First, I would like to thank Mr. E. M. Buchan and Mr. W. H. Westphal, Managing Directors of Rentokil Limited, for permission to use statistical and other information relating to the woodworm problem. The publication of such information concerning commercial practices is of the greatest value and is an example of their farsightedness.

Many of my colleagues of the Research Laboratory and Organizational Staff of Rentokil Limited have given me great help. I am particularly grateful to Miss D. Linscott for the painstaking way in which she reads my typescripts and proofs and for her very helpful suggestions. Mr. Alan Farrington, Group Marketing Manager, has studied the typescript and his valuable criticisms have been incorporated in the present text.

Many of the photographic illustrations are the work of Mrs. E. R. Winslow of the Rentokil Research Laboratory and I am very grateful to her for this service.

In addition, I wish to thank the late Mr. Rex Beeching of the Woodworm Insurance Company for his valuable help.

Mr. W. E. Bruce of the British Wood Preserving Association has given advice on the content of the book and his kindly criticism is much appreciated.

INTRODUCTION

Wood is a very important material in our national economy. It is the most versatile substance used by man and has played an exceptional part in the evolution of his civilization. Perhaps the dawn of civilization was when fire was first controlled—maybe at the edge of a smouldering forest fire. Certainly, burning for warmth or cooking is still one of its most important uses today. But since the earliest times man has fashioned wood for almost every conceivable purpose—weapons and tools, boats and wheeled carts, musical instruments and buildings.

Today an immense amount of wood is in use in Britain. A very high proportion of the 16,000,000 houses are of partly timber-framed construction and in addition a high proportion of the domestic furniture in use today is made of wood. These two basic uses of wood are the main subject of this present account. It is not generally realized that the annual value of wood imported into Britain is of the order of £350 million—say £1 million each day. Home-grown timber is about 10% of this figure, but due to the maturation of forests it is likely to increase substantially within the next ten to twenty-five years, but most of our timber arrives at our ports after a long and expensive journey and, moreover, due to several other causes, timber is becoming increasingly expensive. It seems a logical conclusion, therefore, to look to the conservation of timber to make it last longer. This is of great importance because, as will be explained in a following chapter, it seems that the more wood we use, the shorter time it is likely to last, *unless we do something about it.*

1

WHAT CAUSES WOOD TO DECAY

Wood-rotting fungi and wood-boring animals—What is woodworm?—The relative importance of the different species of wood-boring beetles—The incidence of wood-worm—Where does woodworm come from?

Before discussing the special problem of woodworm we should first of all consider the factors and agencies causing the destruction or decay of wood and thus see where wood-boring insects enter the general picture. Wood is an organic substance—the woody tissue elaborated by certain plants—and, in common with all other substances which are the product of living pro-cesses, can be consumed by other living organisms which cause it to deteriorate. It should be thoroughly understood that 'timber does not deteriorate through age alone'. These factors or agencies may be vegetable, animal, mineral or physical and can be summarized as follows:

Factors and Agencies Causing Destruction or Decay of Wood

1. *Attack by fungi:* Fungi are plants, members of the vegetable kingdom. They differ, however, from all other plants by not possessing the green colouring matter chlorophyll, with the aid of which energy from sunlight is absorbed to convert the gas, carbon dioxide, and water to sugar and starch. Fungi then must be either parasitic—deriving their nourishment from living organic tissue such as the fungi often present on living trees, or saprophytic—deriving their nourishment from organic tissue already dead or decaying. The Dry Rot Fungus—*Merulius lacrymans*—would be an example of this type.

The fungal hyphae—the minute delicate threads which rot the wood by feeding on a constituent of it (the cellulose)—can only be seen after sectioning, staining and mounting under the high-power objective of the microscope, but certain 'gross' features of fungal attack are always present when such decay has proceeded for some time. The wood becomes light brownish in colour and cracks, either transversely (across the grain) (Fig. 1), or longitudinally. If the wood is painted or otherwise decorated, then severe buckling usually takes place. Considerable weight is lost and a 'sporophore' or fruiting body which produces the spores may ultimately develop.

The more obvious wood-rotting fungi, such as the Dry Rot Fungus *Merulius lacrymans*, belong to a group known as the BASIDIOMYCETES, but in recent years attention has been drawn to a group of primitive fungi

11

Fig. 1. Wood showing attack by the Dry Rot Fungus—*Merulius lacrymans*. Note cracking into cubes and patches of white mycelium.

exemplified by the species *Chaetomium globosum* which causes a superficial rot on the surface of wood known as 'Soft Rot'. This type of rot is known to favour the establishment of several groups of wood-boring animals. The presence of other types of fungal decay is essential for the attack by several species of wood-boring beetles, described in greater detail in later chapters.

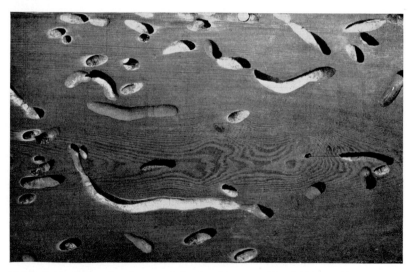

Fig. 2. Sawn timber showing attack by the shipworm, *Teredo*. Note white calcareous lining to the tunnels. This is a valuable character for identifying this type of attack.

Fig. 3. Mechanical wear is another factor causing decay of wood. Probably mainly caused by abrasion, innumerable tiny cuts causing the timber to wear away. The illustration shows a wooden 'pebble'. A block of wood has had all the corners worn off by the action of waves washing it up and down the sea shore.

Fungal decay is extremely important in wood used out-of-doors whereas indoors it is much less important than woodworm attack. Fungal decay requires a moisture content in the wood greater than that normally found in buildings except when there is accidental leakage, faulty building maintenance or in old buildings lacking a damp-proof course. No comprehensive account of fungal decay is given in this book as such accounts already exist. The explanation of fungal decay which is given here will be sufficient for the reader to understand this phenomenon. References to books giving fuller accounts of fungal decay are given in the list of references at the end of this book.

2. *Attack by wood-boring animals:* This section includes insects such as beetles (COLEOPTERA)—the beetles whose larvae are known as woodworm

Fig. 4. Wood decayed by chemical action. An alkali, sodium carbonate solution, has caused the destruction of this wood. (Crown copyright. Reproduced by permission of the Director, Forest Products Research Laboratory).

are placed here, wood-boring wasps (HYMENOPTERA), termites (ISOP-TERA), etc., CRUSTACEA such as *Limnoria*, known as the 'Gribble', MOLLUSCA such as *Teredo*, known as 'Shipworm'. The Gribble and the Shipworm attack only wood floating or submerged in the sea and are known as the marine borers. They were, and still are to a much more limited extent, grave hazards to wooden-hulled ships.

Fig. 5. Decay caused by physical factors. Sample of timber showing prolonged exposure to moderate heat giving appearance not unlike that due to some types of fungal attack.

Temperatures well below that of boiling water if sustained for a long period often cause such decay. Warm water decay is sometimes associated with fungal attack of a special kind.

3. *Mechanical wear:* The effects of abrasion can be observed on well-used unprotected wooden stairs. Often sliding drawers in furniture abrade badly, throwing off a fine dust often confused with the bore-dust of woodworm.

4. *Chemical degradation or decomposition of wood:* Many chemicals have the effect of destroying or weakening wood to a serious extent. Acids and alkalis have this effect. Often wood in contact with acids for a prolonged period has a woolly appearance, due to the separation of the wood fibres, (Fig. 4). One phenomenon rightly belonging to this category is the inter-action of certain timbers on iron nails. The complete rusting of iron nails in oak and other timbers with a simultaneous cracking of the wood is well known.

5. *Heat:* Whilst prolonged low temperature has little or no effect on wood (provided the moisture content of the wood is not unduly high), prolonged heating even at temperatures below that of boiling water causes breakdown of the wood—often simulating fungal attack to a remarkable degree, (Fig. 5).

6. *Bacteria:* Bacteria play an almost negligible rôle in decay of wood, yet their presence has been shown in certain situations. The presence of anaero-bic (living in the absence of oxygen) bacteria decaying the outer skin of

softwood piles lying many feet in wet silt under the City of Rotterdam has been demonstrated. The bacteria-decayed wood is similar to fungal attack. It is discoloured to a greyish-brown and there are innumerable fine transverse cracks, (Fig. 6).

In each category of the summary, some timbers withstand the agencies of destruction to a greater extent than others. In other words some timbers are more *durable* than others, with particular factors in mind. It does not follow that a species of timber is durable against all possible factors.

Fig. 6. Bacteria cause extremely little decay of wood but we illustrate one of the rare instances—a piece of pine piling put into position under the City of Rotterdam in the sixteenth century and blown out during the bombing in 1940 from a position eighteen feet deep where it lay in wet mud. The bacteria causing this decay are of the anaerobic type—that is, they live and decay the timber in the absence of oxygen.

What is Woodworm?

Woodworm is the name given to the larval or immature stage of certain wood-boring beetles found in buildings. Strictly speaking, then, they are not worms but are analogous to the grub or caterpillar stage of a moth or butterfly. Until some forty years ago the term 'woodworm' meant only the larvae of the Common Furniture Beetle (*Anobium punctatum*) or Death-Watch (*Xestobium rufovillosum*) as the various *Lyctus* species were only then becoming common after introduction from countries abroad. It is interesting to note that in New Zealand the Common Furniture Beetle (*Anobium punctatum*), which is exactly the same species as found in Britain, is known by the much more applicable name of 'House-borer'. This is in view of the fact that for every furniture beetle larva tunnelling in furniture there must be thousands tunnelling in the joists, rafters and floors of houses.

What are the gross features of a woodworm infestation; that is, what are the naked eye signs and how may it be distinguished from fungal attack?

Normally, the presence of flight holes, usually circular, is the first sign of woodworm infestation. The adult or beetle stage, if in sufficient numbers, may also be observed, but it is amazing how often beetles may be present in hundreds, yet the householder is unaware of their existence, or even if they are observed, their connection with woodworm is not inferred. The presence of 'dust' or wood powder is often associated with the flight or exit holes, very much more so in the various species of *Lyctus* than Furniture Beetle and other species. This is brought about by new boring intercommunicating with an old tunnel leading to a flight hole. It is important to remember that the powder is kicked by a larva out of a flight hole through which it will *not* itself emerge. Sometimes powdering is caused by predacious insects which are feeding on the woodworms.

An insect is an animal distinguished from all other animals by having six jointed legs in the adult stage, with a segmented body. During its life cycle it undergoes a complete metamorphosis (that is, the change from grub or larva through pupa or chrysalis to adult), or the metamorphosis may not be complete—the young individuals looking very much like the adults except in the possession of wings. Wood-boring beetles are examples of the former type with a complete metamorphosis. Regular circular tunnels in the wood, flight holes on the exterior surface, the associated grub-like larvae inside the wood and the resulting beetles outside the wood are all characteristic of woodworm damage, easily distinguished from fungal decay described previously.

Which Species Does Most Damage?

What is the relative importance of the various species of wood-boring beetles? If we assess the damage caused by wood-boring beetles in the United Kingdom—damage caused to all fabricated woodwork such as furniture, structural timber, floorings and indeed all the things that man has fashioned out of wood—and then assess the proportion of the whole caused by each species of wood-boring beetle, what sort of result would we get? A truly scientific approach to this question at the present time would be very difficult. What units would we employ? The number or population of various insects would obviously not be an accurate guide. It would take several hundred furniture beetle larvae to do the same damage as one House Longhorn larva. Again there are quite local areas where one species predominates—such as an area of Surrey where House Longhorn occurs—but when viewed against the country as a whole, a very different perspective is obtained.

Another difficulty in assessing both the relative importance and the actual incidence of woodworm attack, if the word 'attack' or 'infestation' is used, is to say what constitutes an infestation—one flight hole or a thousand? Obviously agreement or a rule regarding this must be obtained by the authorities concerned.

Although the relative importance of the various wood-boring beetles has been known approximately for ten years, during the last two or three years the results of considerable numbers of timber surveys in buildings carried out by a large woodworm-exterminating company and by a woodworm-insurance company have been published. The relative importance, therefore, is now known accurately (actually previous estimated figures were found to be fairly accurate) and in addition variations in the distribution of the different species in the various parts of the United Kingdom are shown. The figures show in a graphic way the various proportions of wood-boring pests according to the statistics gathered from the first-mentioned company. During 1962, the timber in 21,303 buildings was examined and infestations by wood-boring insects and wood-rotting fungi reported upon. It should be stated that the owners of all these buildings had requested the company to make the survey so that it must be realized that the 'sample' of buildings was not a 'random' sample, as many of the owners must have at least known that the condition of the timber in the buildings was suspect. As far, however, as the comparative incidence of wood-boring insects is concerned, this does not matter. It is of interest to note how this sample is made up in respect of types of buildings, which were as follows: 84·8% private dwelling houses, 6·2% commercial premises, 4·8% churches and historic buildings, 0·9% licensed premises and 3·3% farms.

The age distribution of the buildings in the sample is also of interest and was as follows: pre-1919 52·1%, 1920–39 41·9%, and 1940–62 5·1%. The percentage of the various groupings did not vary much throughout the United Kingdom except that in Scotland it was: pre-1919 69·5%, 1920–39 26·4% and 1940–62 3·5%.

The Common Furniture Beetle

The important result of the survey, which surprises many people, is the great preponderance of damage caused by the Common Furniture Beetle (*Anobium punctatum*). 75·4% of the buildings surveyed showed sufficient woodworm damage by this insect as to merit control by a servicing company. Throughout the United Kingdom the comparative importance of *Anobium punctatum* is uniformly high but is highest in the English southern counties, the figure there being 84·3% and in the south-western counties, where it is 81·1%. It was found to be lowest in the north-western counties at 63·0%, whilst in Scotland it was 71·6%. Quite independently of these results it was found in 1962 by the officers of a local authority that 85% of pre-1939 houses in a council estate in Shoreham, Sussex, had 'Woodworm'.

Death-Watch Beetle

The number of infestations attributed to Death-Watch amounted to 907 or 4·3% of all the buildings surveyed, and this was the next most important wood-boring beetle. It was not found in Scotland nor in Northern Ireland. This was to be expected, as this insect is extremely rare in those countries.

B

Fig. 7. The Life-cycle of Common Furniture Beetle: *Anobium punctatum* de Geer. This shows an insect with complete metamorphosis, i.e. going through a larval and pupal stage, the larval stage being very different in outward appearance from that of the adult beetle. The life-cycle is passed through in about three years but it is the 'worm' or grub stage which is responsible for most of the damage, although the adult beetle biting its way out is responsible for the little holes, the only visible damage.

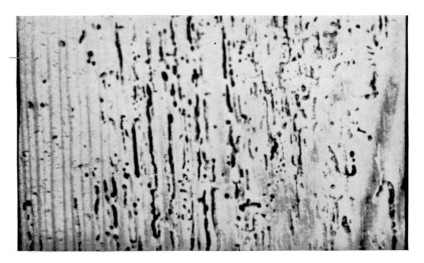

Fig. 8. Deal floorboard with the surface planed to show the tunnelling inside by *Anobium punctatum*. Damage equal to this can seriously impair the load-carrying properties.

Fig. 9. An attack of Common Furniture Beetle out of doors in the branch scar of an old apple tree.

Lyctus or Powder-Post Beetle

Lyctus beetle attack was found only in 0·9% of the buildings examined. At one time it was thought to be a much more serious problem. In the north of England and Scotland the figure was only about half that given.

The Wood-boring Weevils

Attack by the two species *Euophryum confine* and *Pentarthrum huttoni* was found in 4·5% of all buildings but it was abundant in London, occurring in 11·8% of buildings. It occurred only twice in Scotland. Exact identifications were not made as it is difficult in the field, but the former species is thought to dominate.

Fig. 10. Common Furniture Beetle showing position of open wing-cases and wings ready for flight.

House Longhorn Beetle   *Hylotrupes bajulus*

It is probable that the full extent of damage due to this insect within the area of its distribution is even now not appreciated. This is due to the habit of House Longhorn of preferring the micro-climate of the roof spaces of houses in which to carry out its life cycle, and thus often escapes detection. In the surveys during 1962, 90 infestations were found, mainly in Surrey and Hants, being 2·5% of all buildings examined in that area, but in the country as a whole it was found in only 0·4% of all buildings.

A large number of beetles share the responsibility for the last small fraction of damage. A few of the more common ones are dealt with later.

The Incidence of Woodworm

The 'true' or 'actual' incidence of woodworm is, of course, something quite different from the comparative importance of the different species. We are now considering the total number of infested buildings in the United Kingdom. A survey of 16,000,000 buildings would be a colossal task to perform but, as carried out in many other types of survey, a cross-section or sample can be examined and if our sample is truly representative we can infer that what we find in the sample bears the same proportion in the whole.

There are many problems in selecting buildings for the sample survey in making sure that the proportions of the various types of buildings in the sample are fully representative in all respects that are likely to influence, one way or another, the incidence of the various wood-boring insects. A thorough study must be made of all the factors, biological and social, before an authoritative survey can be made. On the other hand there are certain indications which point to such a substantial proportion of Furniture Beetle incidence that mention should be made of it here. During part of 1961, 694 timber surveys were carried out in respect of applications for insurance against woodworm. There is reason to believe that the owners of a large proportion of these properties were under the impression that woodworm did not occur in their homes. It is of interest to note the age groupings of this sample of buildings. 26·2% were built before 1919, 47·9% were built between 1920 and 1939 and 13·9% built between 1940 and 1962. Of the 694 buildings, 415 or 59·8% contained an active infestation of *Anobium punctatum*. Of the 166 buildings surveyed that were built between 1940 and 1962—what we can term *new* houses—36, or 21·7% were infested with *Anobium punctatum*. At the present stage of our study of the nature and extent of the statistics available, it seems certain to the writer that at least half the buildings in the United Kingdom contain an active infestation by the wood-boring beetle *Anobium punctatum*, and of the houses built since the end of the last war it is probable that something like one in five is infested.

Where does it come from?

Where does woodworm come from? Has woodworm a habitat outside man's economy? How does woodworm enter a house in the first place? Can it fly? Here are questions which everyone asks when woodworm of one species or another enters his own home.

Now Furniture Beetle, *Lyctus* and Death-Watch are found quite commonly out-of-doors. Indeed Death-Watch is much more frequently met with outside than within. Many hardwood trees—oak, elm, ash and especially willow—which are dead, perhaps struck by lightning, or with some dead part such as a branch or wound where a branch has been lopped off, will harbour some Death-Watch. Old pollard willows are perhaps the most fruitful source of these insects and the dead wood in the crown of the tree always seems to support a good colony. Death-Watch is found in such situations where fungal decay is evident, whereas Furniture Beetle, on the other hand, is seen to occur where for some reason the wood has resisted the onset of fungal attack, possibly remaining fairly dry by sheltered position or good drainage. Old branch scars on trunks of trees or areas of exposed sapwood are almost always populated by Common Furniture Beetle, (Fig. 9). The freedom from fungal decay in such situations where this insect is established leads the writer to conjecture whether some fungicidal material is secreted into the timber by the developing larvae.

Lyctus, too, occurs in timber out-of-doors. It attacks, however, only the sapwood of certain hardwoods such as oak, which have been newly felled, ringed by animals, uprooted by storm, or where for some other reason the sap flow has ceased. Only for a few years does *Lyctus* maintain itself on such timber—only while the starch content in the sapwood is high. Only perfectly sound timber is attacked. *Lyctus* is never found in timber already attacked by fungus. It is most commonly found in timber yards.

We have thus dealt with the question of where woodworm occurs outside our homes—at least with the three most common species. Now we must see how they get into our homes. With Common Furniture Beetle and *Lyctus* the explanation is simple. The adult beetle possesses a good pair of wings tucked underneath the wing-cases (the elytra) and both species fly readily. Common Furniture Beetle in flight may often be mistaken for a house fly, even by the experienced entomologist, so closely does it resemble the latter, even to flying round central light-coloured objects in a room, such as electric light bulbs.

Beetles that have made a flight can nearly always be identified as they are not usually able to fold the tips of their wings under the wing-cases once they have flown. Furniture Beetle is often met with out-of-doors and beetles may settle on a person's clothing and be brought indoors. It can, of course, fly through the open window in either direction. As far as flight is concerned *Lyctus* is very similar in habit to Common Furniture Beetle. It does seem, however, that *Lyctus* is attracted to light to a greater degree than Common Furniture Beetle and the adult beetle may often be seen in the flight season on the window panes where an infestation occurs in a room. It is not known how far these two species can fly but their flight is quite strong enough for them to be carried considerable distances if caught by the wind.

Death-Watch Beetle cannot fly—it is known only to steady itself by lifting the elytra—and it seems certain that this insect is introduced into a building with the original timbers. The latter were invariably of large section and cut from over mature oak and chestnut trees already infested with a wood-rotting fungus and containing the Death-Watch larvae.

Notes on the Scientific Names of Insects

The reader will have observed in going through this chapter that often when the common (English) name of an insect or plant is given it is followed by the Latin, or latinized words, always printed in italics. You will want to know the reason for this, which is as follows: The common names of our animals and plants differ in the various localities and in addition would be quite unknown to foreign biologists. It would be quite unusual for the common names of insects to be found in English dictionaries! Biologists have therefore evolved a system of naming animals and plants which has been universally adopted and is known as the Linnaean system (named after its founder, Linnaeus—1707–78). These rules of naming are rather complicated but very exact, and although anomalies do arise from time to time, taken generally the system works very well. Expressed very simply it is as follows.

The natural classification is the basis of the system. That is, all the animals and plants are arranged in the way that is thought that they evolved and names are given to the important groups that have characters in common. For instance, the 'class' INSECTA is divided into a number of 'orders', examples being COLEOPTERA (beetles), LEPIDOPTERA (butterflies and moths) and DIPTERA (two-winged flies). The orders are again divided into families and the families sub-divided into the smallest groups, which are called genera. Each one of these latter is called a genus. The individual species are grouped into the genera. Each genus is given a name consisting of a single word and each sort of individual likewise. The name given to a genus is never duplicated, but this may be done in the case of the individual name. Thus, the combination of the name of the genus and the name of the individual sort means one particular animal, and furthermore its relationship with a group of animals will be seen at once. The generic name is always commenced with a capital letter and the individual or trivial name as it is called, is always commenced with a small letter. Both words are always underlined in manuscript and typescript to show that, if printed, the whole name of the insect would appear in italics. More correctly the name of the author (who first described the species and named it) should be given afterwards, either in full or in an abbreviated form, but this rule need not be adopted on every occasion.

2

COMMON FURNITURE BEETLE

How to identify it—Woods liable to attack—Parasites and Predators

The great importance of the Common Furniture Beetle has already been stated—its preponderance in wood-boring insect damage in the United Kingdom. Here is given a description of the biology of this beetle and the feeding habits and the timber infested are next discussed.

Common Furniture Beetle is the colloquial name given to the insect which de Geer called *Anobium punctatum* in 1774. (*Anobium punctatum* de Geer is referred to in literature as *Anobium striatum*; this was the name given to the insect by Olivier in 1790; and also as *Anobium domesticum* given by Geoffroy in 1785. In accordance with the law of priority these latter names are now not used.)

In those days and indeed until relatively recent years *Anobium punctatum* was a beetle associated with old houses and old furniture. For instance, Stephens in 1839 said that this beetle was 'abundant in old houses through-out the country' whilst in 1931 Joy, in his *Practical Handbook of British Beetles,* said 'the larvae attack furniture which is said to be "worm-eaten"'. Today the situation is very different. Only some thirty years since Joy wrote this, as has already been stated, half the houses in the British Isles are probably attacked to some degree or another by this beetle. The writer's theories to account for this increase are given later, after the section dealing with the life-cycle and the types of woods attacked. They will then be better understood.

Common Furniture Beetle belongs to the family ANOBIIDAE which contains about twenty-two different species found in Britain. This total includes the Death-Watch Beetle and several other species of economic importance. Among these is *Stegobium paniceum,* often known as the Biscuit Beetle, which is a common pest of stored foods, soup powders, etc. (It is often mistaken for the Furniture Beetle, as it has a superficial resemblance to this species.)

The adult beetle varies in size from one-tenth to nearly a quarter of an inch in length and it is usually chocolate-brown in colour, while lighter and darker individuals are often encountered. All are covered with a fine yellowish down which does not, however, hide the longitudinal rows of 'punctures' or small pits on the 'elytra' or wing-cases. The most charac-teristic feature, however, of this beetle is the hooding of the head by the prothorax or the chest. This is shown in the illustration of the side view of the beetle, (Fig. 11). To a greater or lesser extent, this hooding of the head is shown by all members of the family ANOBIIDAE as well as by members of the related families PTINIDAE (the Spider Beetles) and BOSTRICHIDAE (the Powder-Post Beetles).

From twenty to sixty eggs are laid by each fertilized female *Anobium*. The maximum possible is about eighty but the average is said to be twenty-eight. They are laid usually in small groups, in cracks of unpolished wood,

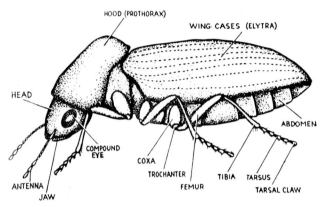

Fig. 11. Side view of Common Furniture Beetle showing the hood and the longitudinal rows of 'punctures' on the wing-cases.

(Fig. 12), but sometimes they may be found just inside the old flight holes. In shape they are like oval pearls, but they are often pushed a little out of shape when being forced into crevices during egg deposition. They are easily seen by the naked eye and the illustration shows a group of eggs found on the plywood back of a picture frame. They are seen to be wedged underneath the rough splintery grain.

The young larvae emerge in about five weeks' time, (Fig. 14) not from the top of the egg but from the base, and immediately commence to tunnel their way into the wood (compare with the young larvae of Death-Watch). The larval stage is spent entirely in the wood and extends for not less than two years, the average length of the larval stage probably being something between three and four years; periods of five or even more are perhaps not uncommon. Out-of-doors an annual life cycle is said to take place.

The larva is greyish-white in colour and covered with fine hairs. The head is yellowish-brown, but the jaws are dark chestnut-brown. It usually holds its body in a hook-shape when in its tunnel, or tightly curved into a ball when it is brought out of the tunnel into the light.

The larvae remain entirely within the wood, tunnelling mostly up and down the grain, but often crossing into a different growth ring. Tunnelling takes place in cold weather as well as in summer, although at a decreased rate, and after tunnelling the larva withdraws a short distance from the 'working face' for an hour or two before recommencing operations. The tunnel which is left behind is loosely filled with bore-dust or 'powder', which is known as frass. The powder when rubbed into the palm of the hand is gritty, quite distinguishable from the powder of *Lyctus*, which is soft and silky, (Fig. 31). If the powder of *Anobium* is viewed under a lens it is seen to consist of cigar-shaped pellets made up of minute fragments of

Fig. 12. Eggs of Common Furniture Beetle, seen here wedged in a crevice of birch plywood. Much enlarged.

the chewed wood. In severe infestations where the galleries intercommunicate large amounts of the frass may be forced out of the old flight holes. 'Powdering', however, does not occur to the same extent as in *Lyctus*.

In the early spring of the year in which the larva matures, it bores towards the outside of the wood, but just short of the actual surface it constructs a pupal chamber slightly larger than the diameter of the gallery and changes into a pupa or chrysalis. The pupa is creamy-white in colour and in many ways resembles the adult beetle. All the legs, wing-cases, antennae, etc., can be seen plainly, but are held quiescent by the thin transparent pupal skin. It can move only the last few segments of the abdomen, but in about six to eight weeks the pupal skin is burst off and the adult eclodes.

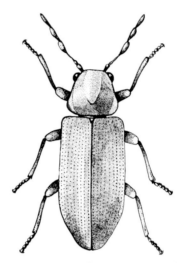

Fig. 13. Common Furniture Beetle—view from the top. It is about one-eighth of an inch in length and dark chocolate-brown in colour.

Although the adult stage has at last been reached a period is spent resting whilst the parts harden and turn to the characteristic brown colour. Then the beetle bites its way out into the open air and the hole it makes, which averages about one-sixteenth of an inch in diameter, is termed the 'flight hole'. The adult beetle stage extends over two or three weeks only. During this time mating takes place (sometimes within a few hours of emergence) and then the female lays her eggs in the sort of places described and illustrated at the beginning of this chapter. It is of interest, however, to mention that the females will stay in old flight holes, both before mating and afterwards, and the act of mating often takes place with the female beetle just inside the old flight hole with the tip of her abdomen just protruding and the male on the outside of the piece of wood. If insecticidal treatment of the piece of wood is carried out when a number of females are in hiding,

Fig. 14. Approximate average length of individual life stages. Left, Common Furniture Beetle. Right, *Lyctus*—see Chapter 4. This illustrates the much faster rate of the life-cycle of *Lyctus* compared with that of the Common Furniture Beetle. In both cases only the larva feeds and grows.

they back out of the holes greatly disturbed, but the layman often gathers the impression that the fluid has caused them to bite their way out of the wood, which, of course, they have not done.

Many wood-boring insects (and *Anobium punctatum* is one of them) have the power to bore through materials other than wood when endeavouring to escape from it on the assumption of the adult stage. Escape from painted and french-polished wood is, of course, very common and well known. In this case, if all the free surfaces of the wood are well painted then the process of painting must have taken place subsequent to the boring of young larvae into the wood. Linoleum and leather are two materials ofter perforated by the emerging adult Furniture Beetle—the latter without difficulty, but linoleum does appear to kill a proportion of them as will be seen by an examination of the underside of perforated linoleum, when some dead beetles will usually be seen imprisoned in it. A variety of made-up goods lying in warehouses as well as books in libraries have been tunnelled into by emerging adults making their way out of wooden shelving.

Woods Liable to Attack

At this point no apology is made for giving some brief notes about the botany of wood, because without some elementary knowledge of timbers

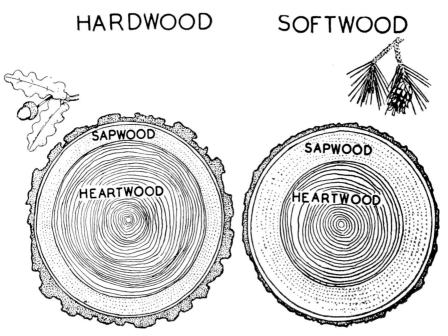

Fig. 15. Transverse sections of the trunk of a broadleaved tree (left) and a coniferous tree (right). The sapwood in each case contains living tissue but the heartwood consists only of dead tissue.

the significant differences between the biologies of the different wood-boring beetles would not be understood.

In the first place we have to distinguish between the timbers of the cone-bearing trees (Gymnosperms) and the broad-leaved trees (Angiosperms). The timbers of the former trees are known as softwoods, and the latter hardwoods. (Examples of softwoods are fir, pine, spruce, larch, and of hardwoods oak, elm, ash, birch, mahogany, walnut, obeche). It is important to understand this definition because today many tropical hardwoods which are now coming into large-scale use are actually softer than many softwoods and on the other hand there are several 'hard' softwoods.

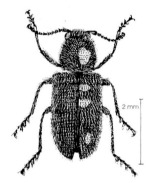

Fig. 16. The adult beetle of *Korynetes caeruleus*.
This beneficial insect is steely-blue in colour.

In both types of timber, however, the wood is differentiated into the outer layer of sapwood and the inner core of heartwood. This is shown in the accompanying illustration, (Fig. 15). In many timbers there is a great difference between heart and sap (such as laburnum where the heart is chocolate-brown and the sap canary yellow) but yet in others heart and sap may be identified only with difficulty. There is, however, this great biological significance; only the sapwood contains living tissue and only in this outer layer of wood is water transported throughout the trunk and branches. The heartwood consists of dead tissue only and into the heartwood is often excreted matter no longer required by the tree for the living processes. It is often this deposition of waste matter which alters the colour or the texture of the heartwood (and, of course, its durability and many other properties).

The subject of the types of wood which *Anobium punctatum* will infest is not a simple one, indeed there are several controversial points, but the following seems fairly certain, at least in the British Isles.

Softwoods. Historically, it seemed that only old and well-seasoned wood was attacked; then about fifteen years ago the author found that the sapwood of softwood was attacked, but usually not before the timber had been converted some twenty years. The heartwood of softwoods was attacked some time afterwards. Today, however, softwood is reported as being attacked by *Anobium punctatum* long before this time, although it appears that in such cases the attack proceeds at a fairly slow rate for a few years.

It does appear to the writer that the possibility of a genetic change taking place in the insect cannot be ruled out. Just as resistance to the action of chlorinated hydrocarbon insecticides has been bred into certain insect species, so the use of certain softwood species, perhaps from North America, where *Anobium punctatum* is not indigenous, is producing a race of this insect which thrives on new softwood.

Hardwoods. Here again the wood appears to require a certain degree of maturation before attack takes place, but this period is in some cases greater than is the case with softwoods. This has been generally recognized for a long time in the layman's mind—the presence of a few flight holes would provide ample proof of the genuineness of a piece of antique furniture. Beech appears to require a period of about forty years, and the large number of pianos of German origin manufactured between forty and sixty years ago, containing substantial quantities of beech and now infested with Furniture Beetle, is an example of this.

Some timbers are, however, anomalous in this respect. Birch is an example. Plywood made from birch is often found to be infested within four or five years of manufacture. Similar to birch in this respect are the timbers of alder and willow.

Almost all timbers are capable of being infested by *Anobium punctatum*

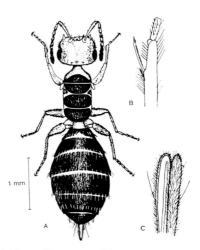

Fig. 17. The small blackish ant-like pteromalid wasp, *Theocolax formiciformis*, is a beneficial insect, its larvae being parasitic on the larvae of Common Furniture Beetle.

but it has been observed by the writer that all timbers carrying a live infestation are perfectly sound, that is without apparent fungal attack.

Let us now examine the increase of Furniture Beetles. If we go back at least twenty years we find the susceptible softwood timbers, and there seems to be no doubt that the increase is related to the vast amount of softwood which came into this country between 1920 and 1940 for the housing programmes. All this wood is now susceptible to attack and the laws of probability have thus made it very much easier for a female Furniture Beetle to find suitable wood for egg-laying. A few years ago Furniture Beetle seemed to be almost a museum piece, but the vast social changes in housing have tipped the balance in favour of this pest.

The following home-grown hardwoods are attacked by Common Furniture Beetle but those timber species where no information is available are also listed:

Alder	*Alnus glutinosa*	Oak pedunculate	*Quercus robur*
Apple	*Malus pumila*	Oak sessile	*Quercus petraea*
Ash	*Fraxinus excelsior*	Evergreen Oak	*Quercus ilex**
Aspen	*Populus tremula*	Turkey Oak	*Quercus cerris**
Beech	*Fagus sylvatica*	Pear	*Pyrus communis*
Silver Birch	*Betula pendula*	London Plane	*Platanus acerifolia*
White Birch	*Betula pubescens*	Black Italian	
Box	*Buxus*	Poplar	*Populus serotina*
	sempervirens*	Black Poplar	*Populus nigra*
Cherry	*Prunus avium*	Grey Poplar	*Populus canescens*
Horse Chestnut	*Aesculus*	White Poplar	*Populus alba*
	hippocastanum	Robinia	*Robinia pseudoacacia**
Sweet Chestnut	*Castanea sativa*	Sycamore	*Acer pseudoplatanus*
Common Elm	*Ulmus procera*	Walnut	*Juglans regia*
Dutch Elm	*Ulmus hollandica*	Cricket Bat	*Salix alba* var
Wych Elm	*Ulmus glabra*	Willow	*caerulea*
Holly	*Ilex aquifolium*	Crack Willow	*Salix fragilis*
Hornbeam	*Carpinus betulus*	White Willow	*Salix viridis*
Lime	*Tilia vulgaris*		

* No information

Some authorities have stated that resin-bonded plywoods are not susceptible to attack. This is a generalization which is not true in particular cases. The writer has seen a bedroom suite in which all the resin-bonded plywood with which it was backed had been infested with *Anobium punctatum*. The explanation is this: before the sheets of plywood were separated from each other, eggs of *Anobium punctatum* were laid around the edge. The resulting larvae thus had two plies in which to develop which were not bonded but just closely pressed to each other. Where the larvae came into contact with the resin bond they turned back away from it. When the sheets of ply were separated the larvae would have fallen out.

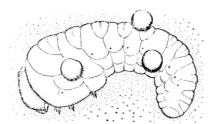

Fig. 18. Larva of Common Furniture Beetle infested
with three *Pyemotes* (*Pediculoides*) mites.

The following home-grown softwoods are attacked by Common
Furniture Beetle:

Atlas Cedar	*Cedrus atlantica*	Giant Fir	*Abies grandis**
Cedar of Lebanon	*Cedrus libani*	Larch	*Larix decidua*
Western Red Cedar	*Thuja plicata**	Japanese Larch	*Larix kaempferi*
Lawson's Cypress	*Chamaecyparis lawsoniana**	Corsican Pine	*Pinus nigra* var *calabrica*
Deodar	*Cedrus deodara*	Scots Pine	*Pinus sylvestris*
Douglas Fir	*Pseudotsuga taxifolia*	European Spruce	*Picea abies*
		Sitka Spruce	*Picea sitchensis*
Silver Fir	*Abies alba*	Yew	*Taxus baccata*
Noble Fir	*Abies nobilis**		

* No information

Some information is now available concerning a number of imported
hardwood timber species. The following are susceptible to attack by the
Common Furniture Beetle:

American Ash	*Fraxinus* sp.	Magnolia	*Magnolia acuminata*
Balsa	*Ochroma* sp.	Obeche	*Triplochiton scleroxylon*

Some species have been found to be moderately resistant to attack:

African Mahogany	*Khaya* sp.	Keruing	*Dipterocarpus* sp.
Agba	*Gossweilerodendron balsamiferum*	Japanese Oak	*Quercus* sp.
		Ramin	*Gonystylus* sp.
		Red Meranti	*Shorea pauciflora*
Central American Mahogany	*Swietenia* sp.		

The Common Furniture Beetle does not appear to be capable of attacking
the following species which are immune or very resistant:

Abura	*Mitragyna ciliata*	Idigbo	*Terminalia ivorensis*
African Walnut	*Lovoa klaineana*	Iroko	*Chlorophora excelsa*
Afrormosia	*Afrormosia elata*	Sapele	*Entandrophragma cylindricum*

It is very interesting to note and certainly gratifying to the owners of Victorian furniture that Central American Mahogany, used extensively for furniture manufacture in the past, has been found to be moderately resistant to attack.

Wickerwork baskets (willow-twigs—entirely sapwood), broken and thrown into the loft, have very often been a reservoir of infestation for many years. They produce an annual crop of adult Furniture Beetles which have laid their eggs on furniture and structural timbers throughout the house. The transport of infested furniture, suit-cases, baskets and other items is probably a very important factor in the spread of infestations from one property to another.

Apart from the different timbers which are attacked, the question arises as to whether Furniture Beetle prefers any special conditions such as temperature or disturbance. Many years of laboratory research would be required to give the answer to these questions, but the prevalence of infestations of Furniture Beetles in the 'cupboard under the stairs' appears

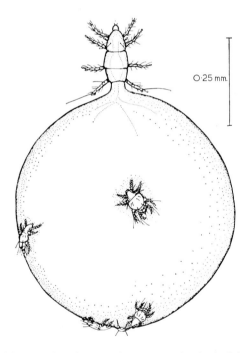

0·25 mm.

Fig. 19. A gorged female mite of the species *Pyemotes* (*Pediculoides*) *ventricosus* Newp. The *'Pediculoides* mite' is of great interest from a biological point of view in that the male mites, which are very small, never leave the surface of the body of their mother. Here they mate with their sisters as they are produced. The young female mites as they are fertilized drop off their mother and seek out a host to parasitize. This may be a Furniture Beetle larva or the larva of some other insect. *Pyemotes* mites probably represent a very large factor in the control of many insect pests. Greatly enlarged.

to the writer not to be a preference for undisturbed situations as such, but to a confined distribution. That is, when once the distribution has commenced (probably due to the birch plywood panelling in the first place), the adult beetles are not able to fly away from the closed cupboard-space and they thus lay their eggs back on to the already-infested woodwork. Thus a considerable infestation is built up relatively quickly. It is also conceivable that the preferred range of moisture content of the infested timber is a little above the normal—say 14 to 18%—a moisture content which would be quite likely in unventilated woodwork.

The Parasites and Predators of Woodworm

Common Furniture Beetle has a number of parasites and quite a few predators preying upon it, some of which exhibit quite extraordinary biologies. It might have been thought that wood-boring insects lived under somewhat sheltered conditions protected by the wood, but such is not the case. Populations of Common Furniture Beetle often decrease, indeed die out altogether, after a period of intense activity, especially when old oak sapwood is being attacked. Whether this is due to parasites, whether the heartwood is not susceptible to attack, or whether it is due to some other cause, is difficult to ascertain, but it is likely that parasites sometimes play a large part in reducing the activity of an *Anobium* infestation. On the other hand it is extremely unlikely that parasites are ever responsible for a complete extermination of woodworm.

A most interesting feature of the present great increase in Furniture Beetle has been the corresponding increase of the hymenopterous parasite *Theocolax formiciformis* Westwood. (The HYMENOPTERA is the insect group containing ants, bees and wasps.) This small pteromalid wasp looks like a small black ant as described in its specific name, (Fig. 17). It is often seen running in and out of the old flight holes of *Anobium* or its own flight holes, which are considerably smaller than the latter, being only about $\frac{1}{64}$th of an inch in diameter. In almost all the adults the wings are rudimentary and even those that possess wings cannot fly. The female lays her eggs (maximum fourteen, average four to eight) on the back of the Furniture Beetle larva or near to it, finding the larva by running through the *Anobium* tunnels or by cutting through the thin layers of wood with the ovipositor. The larval parasites devour their hosts fairly rapidly, completing this in about one to two weeks. The *Theocolax* larvae are described as ectoparasitic (i.e. they live on the surface of the *Anobium* larvae).

The pupal stage lasts about three weeks and the emerging adults make their way to the outside *via* an old *Anobium* flight hole or they may bite their own way out as previously mentioned. Three specimens once gnawed their way through half-inch corks and escaped in the writer's laboratory. The total length of the life cycle varies from about thirty-nine to seventy-five days according to climatic conditions so that emergence may take place throughout the year.

C

It has been found in the laboratory that, when gamma rays had been used to kill *Anobium* larvae, *Theocolax formiciformis* adults emerged apparently unharmed and similarly this species emerged safely from wood which had been subjected to V.H.F. heat treatment sufficient to kill *Anobium* larvae.

Another member of the group HYMENOPTERA parasitizing *Anobium* larvae is the little Braconid *Spathius exarator* (L.), and from 65 to 94% *Anobium* larvae have been reported as being infested with the insect. The adult *Spathius* is often observed walking rapidly over a window pane, obviously strongly attracted to light.

The beetle *Korynetes caeruleus* (de Geer) (Fig. 16), which is quite a common predator of Death-Watch Beetle, also attacks Furniture Beetle. The larva of this beautiful steely-blue beetle moves slowly around the intercommunicating galleries devouring the wood-eating larvae when it finds them. It has something of a superficial similarity to the larva of the Brown House Moth (*Hofmannophila pseudospretella*), but it may be distinguished from the latter by the possession of a pair of blade-like processes at the hinder end of the body, (Fig. 20).

A number of other species of beetle larvae are also recorded as being predaceous on woodworm larvae, amongst which only the rather large *Opilo mollis* deserves mention. It is closely related to the aforementioned beetle *Korynetes caeruleus*, both being included in the beetle family CLERIDAE.

When infestations of *Anobium punctatum* and the Death-Watch Beetle

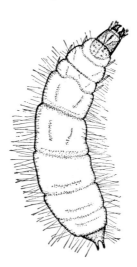

Fig. 20. Predacious larva of the beetle *Korynetes caeruleus* (de Geer). This larva hunts out and feeds on the immature stages of Furniture Beetle and Death-Watch. It is therefore a beneficial insect. The larva grows up to 14 mm. in length, and is creamy-white with a chestnut-coloured head which can be retracted a long way into the first thoracic segment.

have been exterminated or controlled by fumigation or by the use of oil-soluble insecticides it often happens that *Korynetes caeruleus* beetles afterwards emerge unharmed.

Finally, we leave the insect group for perhaps the commonest parasite of Furniture Beetle—the mite (a member of the spider group) *Pyemotes* (*Pediculoides*) *ventricosus* Newp. These are sometimes present in great numbers. They suck the juices of the *Anobium* larvae and the gorged females can be seen looking like shining beads on the larva's body, about one millimetre in diameter, (Fig. 18). The males never leave the outside of the body of their mother, (Fig. 19) and mate with their sisters as they are produced.

All these parasites and predators of woodworm are, of course, beneficial and should be preserved as far as possible. Certainly the expert in wood-worm disinfestation should be able to recognize them. However, it should always be borne in mind that the total extermination of the host (in this case the woodworm larvae) is seldom, if ever, achieved by their help alone, and thus the occurrence of one or more of these parasites should never be a reason for deciding against insecticidal or other control measures.

3

THE DEATH-WATCH BEETLE

Identification and Biology

The Death-Watch Beetle is an indigenous British insect which, on account of its habits, has been well known for several hundred years. It was given the name *Xestobium rufovillosum* by De Geer in 1774, which is the name now generally used, although the name *tessellatum*, given to it by de Villers in 1789, is found in much older literature. Its outdoor habitats consist of dead wood in trees or dead branches of several hardwood species where fungal decay occurs, a common situation being the dead wood in the trunk or crown of pollard willows. Oak, ash and sweet chestnut are also commonly infested by this species, and hornbeam, poplar and whitethorn more rarely.

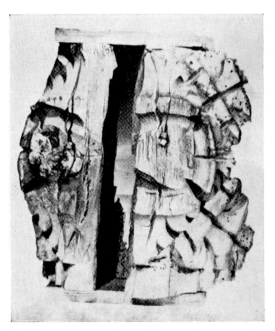

Fig. 21. A boss from Burwell Church, Cambridgeshire, as seen from the ground and showing only slight evidence of Death-Watch attack.

36

Fig. 22. The reverse side of the same boss showing almost complete destruction by the beetle at the back.

Indoors, Death-Watch infests hardwood structural timber which has at one time or another maintained a fungal infection. Oak is the most common timber attacked, but chestnut, elm, alder and walnut are sometimes infested. On occasions, however, softwood is attacked if it is adjacent to infested hardwood but attacks only originate in hardwood.

The Death-Watch infested timber can appear quite sound yet the rich chocolate-brown colour of the tunnelled timber make it very probable that the living tree from which the timber was converted was attacked by the beef-steak fungus, *Fistulina hepatica*. This is the fungus which produces 'brown oak'. In other instances the cellar fungus *Coniophora cerebella*, causing a wet rot, and *Phellinus cryptarum* and *Polystictus versicolor*, causing white rots, have brought about the incipient breakdown of the wood required by the Death-Watch Beetle. In other words, the larva seems to require the wood to be partially predigested for it.

This phenomenon is a common occurrence amongst the wood-boring beetles, but on the other hand the larvae of many of the common species such as Common Furniture Beetle (*Anobium punctatum*), *Lyctus* spp. and House Longhorn (*Hylotrupes bajulus*) feed on sound wood in buildings where fungal infection is apparently absent.

Death-Watch Beetles are incapable of flight; only rarely have they been known to raise their wing-cases just to steady themselves. This raises the question as to how they come to infest timber indoors.

It seems virtually certain that Death-Watch infestations in buildings were initiated by the use of timber of large dimensions already infested when the tree was growing and few new infestations take place today. This means that with the current work of control the total number of infestations indoors is rapidly decreasing.

Life Cycle   *The Egg*

The eggs are laid in small clusters, usually of three or four, on the surface of rough wood, cracks or just inside the lip of flight holes. They are sticky when first laid and adhere to each other. The eggs are whitish, oval and much like those of the Common Furniture Beetle, but are, however, twice as large. Forty to sixty is the average number of eggs laid, but up to 201 have been recorded. The length of this stage varies from about two to five weeks.

The Larva

Again, the larval stage has the same general appearance as in Common Furniture Beetle. A point of difference, however, is found in the eyes. Whereas in Common Furniture Beetle these consist of one black spot on each side of the head, in Death-Watch two black spots occur on each side. Another difference, this time in behaviour, is also apparent. When the young Death-Watch larvae hatch out from the eggs, (Fig. 23) they walk about quite rapidly over the surface of the wood before finding an old flight hole or crevice through which to enter the wood. The larva is creamish-white, strongly hook-shaped when taken out of its tunnel, covered with erect golden hairs and grows to a length of almost half an inch.

The length of the larval stage is very variable. Under the best possible conditions for growth which can be obtained in the laboratory—continued summer warmth, high relative humidity, and with well decayed willow as foodstuff—the life cycle can be completed in a year. Under less favourable conditions it may not be completed until ten years. It is, however, generally much longer out-of-doors: in fungal infected oak it has been found to vary from three to seven years and it is considered that an average of about four and a half years as the larval period indoors would not be unreasonably far from the truth.

The Pupa

This is much like that of *Anobium punctatum*. It is milky-white in colour at first but then darkens progressively.

During the summer in which it reaches full growth, the larva burrows its way towards the outer surface of the wood which it is infesting and constructs a pupal chamber by enlarging the larval gallery and plugging the open end with frass. It changes into the pupal stage in late summer, remaining three or four weeks in this stage, and then changes into the

Fig. 23. Larva of the Death-Watch Beetle leaving its egg. It walks actively over the surface before starting to tunnel into the wood. This contrasts with the larva of the Common Furniture Beetle, which bores through the underside of the egg and tunnels straight into the wood.

adult beetle but remains inside the pupal chamber without biting its way out until Spring. The usual period of emergence is during the latter part of April and the beginning of May. The earliest date recorded is 24 March and the latest 1 August.

The flight holes are circular and approximately one-eighth of an inch in diameter. The pattern of distribution of the flight holes often follows the course of the fungus-infected wood to a remarkable extent.

Identification

The adult beetle is usually just over one-quarter of an inch in length, but may be up to one-third of an inch, the females being usually slightly the larger. It is dark greyish-brown in colour, with a pattern of yellowish scale-like hairs on the pronotum and wing-cases. The wing-cases, however, often become rubbed when the colour may be more reddish and shining. The longitudinal rows of small pits on the wing-cases, present in *Anobium punctatum*, are absent in Death-Watch, (Fig. 24).

The pronotum, which is much more widely flanged, hoods over the head just as in Common Furniture Beetle (Death-Watch Beetle is in the same family—the ANOBIIDAE—as Common Furniture Beetle *Anobium punctatum*).

Fig. 24. Death-Watch Beetle.

The beetle *Korynetes caeruleus* is commonly found preying on Death-Watch. The frass of Death-Watch is quite characteristic; it can be seen clearly with the naked eye to consist of bun-like pellets, (Fig. 31).

The tapping of the adult beetles is a well-known phenomenon. Both sexes tap, and it is the sound of the tapping in the quiet hours of the night in a sick room which has given it its common name.

In captivity the adults can be stimulated to tap by four or five sharp staccato raps with a pencil. The tapping is believed to be of value in finding mates. When the Death-Watch is about to tap, it appears to stiffen and to hold the front of the body away from the wooden surface with the tip of the abdomen touching the surface. It then strikes its head against the wood about seven or eight times within a second, the whole body of the beetle moving.

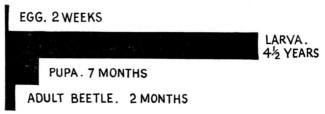

EGG. 2 WEEKS

LARVA. 4½ YEARS

PUPA. 7 MONTHS

ADULT BEETLE. 2 MONTHS

Fig. 25. Death-Watch Beetle, *Xestobium rufovillosum*. Approximate duration of individual life stages. Several months of pupal stage can be replaced by resting stage of adult.

4

LYCTUS—THE POWDER-POST BEETLES

Names of Species—Biology—Significance of starch content in timber—
Timbers infested

Six species of the *Lyctus* genus in the beetle family LYCTIDAE are found in Britain. The names of these species are as follows: *Lyctus brunneus* Stephens 1830; *Lyctus sinensis* Lesne 1911; *Lyctus linearis* (Goeze 1777) (*canaliculatus* Fabricius 1792 and *fuscus* (Linnaeus 1758) are synonyms); *Lyctus planicollis* Leconte 1854; *Lyctus cavicollis* Leconte 1866; *Trogoxylon parallelopipedum* (Melsheimer 1846) (previously included in the genus *Lyctus*). In addition to these six species *Lyctus africanus* is occasionally imported from various parts of the world.

Only the species *L. linearis*, however, ranks as a truly indigenous species. *Lyctus brunneus* was introduced into Britain many years ago from North America, so that today it is common and widespread. The other species, although now established in Britain, have only been introduced in recent years. They are very much less common than *L. brunneus*.

They are all similar in general appearance (usually requiring a trained entomologist to identify the various species) and as their life cycle and habits are also almost identical the following description will serve for all species:

The Egg

The egg of *Lyctus* is long and maggot-like, approximately cylindrical, but tapering off to a tail-like thread. The female beetle places her eggs with care and precision in the early wood vessels or pores, (Fig. 30). These may be exposed in the sectioned wood as cut or indeed opened by the beetle herself in the 'Tasting' marks. They are placed nowhere else, and this is a strong, if mechanical, reason why *Lyctus* damage is restricted to the so-called 'wide-pored' hardwoods. Up to fifty eggs are laid.

The Larva

When the egg hatches, the young larva finds itself already in a pore, but is smaller in diameter than the pore. It then feeds on a yolk-like substance left behind in the egg until its girth is sufficient to enable it to move down the pore by gripping the sides with its body. It makes its way along the pore and finally pierces it. The larva then bores its way backwards and

41

forwards in the sapwood, gradually increasing in size to approximately one-quarter of an inch in length. It is creamy-white, hook-shaped and segmented rather prominently. The head is light creamy-yellow with the jaws dark chestnut. The legs are very small, but a distinctive feature of the *Lyctus* larva is the pair of spiracles or breathing pores on the last abdominal segment, which are large and rusty red in colour.

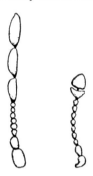

Fig. 26. Antenna of (left) *Anobium* showing club of three segments; (right) antenna of *Lyctus* showing club of two segments only. (Enlarged).

The *Lyctus* larva feeds on the cell contents, containing mostly starch, sugars and related substances together with a little protein. The cell walls cannot be digested. A moisture content in the wood of 8 to 30% is necessary for *Lyctus* attack.

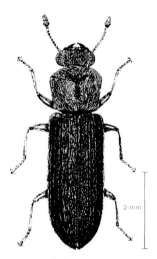

Fig. 27. *Lyctus brunneus.* Note head distinctly visible frcm the top (cf. *Anobium*). *Lyctus* is also longer and narrower than *Anobium punctatum.*

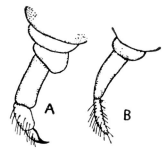

Fig. 28. Leg of larva of (a) *Anobium punctatum* and (b) *Lyctus sp.* Note leg of *Anobium* is five-segmented and terminal segment is clawed, whilst that of *Lyctus* is three segmented with terminal segment paddle-shaped.

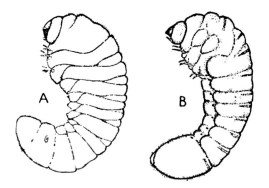

Fig. 29. Larva of (a) *Lyctus sp.* and (b) *Anobium punctatum.* Note that terminal spiracle (or breathing pore) is large in *Lyctus* and small in *Anobium*, whilst head of *Lyctus* is embedded in thorax to a greater degree than in *Anobium*.

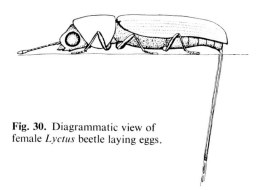

Fig. 30. Diagrammatic view of female *Lyctus* beetle laying eggs.

It has been suggested that a substance other than starch, but which is water-soluble, is the substance which excites the female to lay her eggs, this being the explanation of the fact that fresh-cut timber which has been thoroughly washed in water is immune from *Lyctus* attack.

The conditions under which *Lyctus* attack can take place have been described above. How these delicately balanced biological conditions are satisfied in the practice of felling, extraction, sawing, seasoning and subsequent manufacture of wide-pored hardwoods will be examined next.

The essential part played by the starch content in the initiation of *Lyctus* attack has already been mentioned and it is known that 3% starch content of sapwood of wide-pored hardwoods is required by female *Lyctus* beetles before they will lay their eggs.

Commencing with tree-felling, what changes take place in the starch content of susceptible sapwoods, and in what way is it bound up with moisture content?

When a tree is felled practically half its weight consists of water, and it is, of course, as a consequence, unfit for use for fabricating wooden articles until a great quantity of this water has been removed. Indeed the water content must be brought down to something like 10 to 12% of the dry weight of the wood, otherwise serious difficulties such as splitting and movement will result.

Today this is brought about by one of two methods. The traditional method is known as air-seasoning and consists of cutting the timber into planks of less than three inches in thickness and allowing currents of air to circulate between them by using wooden distance pieces or 'sticks' of certain dimensions. This is carried out in sheds or covered yards and, by skilful manipulation of simple devices, the amount of air circulating is controlled in order to dry the timber with the minimum of faults occurring due to shrinkage. The period taken for air-seasoning was several years, but of recent years the period has been much reduced.

'Respiring' Tissue

Now the starch and other carbohydrate food materials are stored in special tissue in the sapwood known as the Parenchyma. This tissue is living and consequently gradually 'respires' or uses up the carbohydrate material. It will, therefore, be seen that if *Lyctus* occurs at the early stage of seasoning it soon becomes inactive due to the starch depletion, and when the timber is ready for use, those parts of the sapwood damaged by *Lyctus* can be trimmed out and there is no possible chance of *Lyctus* subsequently re-infesting that wood.

Incidentally, softwood 'sticks' should always be employed in the timber yards. *Lyctus* never attacks softwoods, and thus the 'carrying over' of this pest from one lot of timber to another will be avoided.

Within the last thirty years, and accelerated during the last twenty years, the method of kiln-drying of timber has been introduced and today it is practised on a wide scale. There are difficulties in attempting to describe kiln-drying or even air-seasoning within a couple of paragraphs, but this brief summary must be made for an appreciation of *Lyctus* infestation.

Kiln-drying of timber consists essentially of an accelerated moisture-extraction process using steam in large closed rooms or kilns. The steam is used both for heating purposes and for provision of water vapour. In the closed kiln the relative humidity of the atmosphere and the temperature are thus controlled within relatively narrow limits by means of steam valves. A schedule of relative humidity and temperature is used, depending on the species of timber and cut size. The schedule depends on the moisture gradient (i.e. the moisture content at various depths) of the timber being dried in relation to the properties of splitting or bending on drying, but the important point to notice is that this is a moisture-extraction process only and at one time it was thought that the starch content of the timber was little affected. It is now known that everything depends on the initial

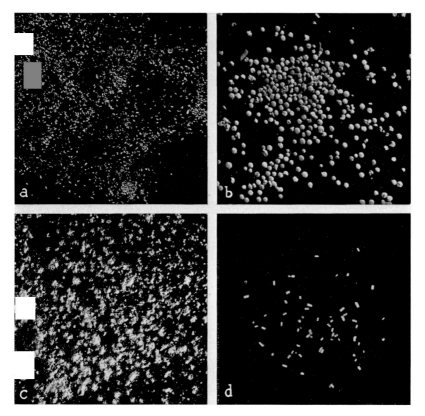

Fig. 31. Larval frass of the following species, much enlarged: (a) Furniture Beetle, (b) Death-Watch, (c) Powder-Post Beetle, (d) House Longhorn.

temperature of the kilning operation. This is critical around 45°C.; below this temperature starch depletion takes place in the kiln (very rapidly around 40°C.), but above it the living tissue is killed and the starch remains. Kilning schedules for oak, ash, chestnut and elm now commence at 35–40·5°C. and therefore the timbers will lose starch. On the other hand kiln schedules for several *Lyctus*-susceptible tropical timbers, iroko, agba and obeche, commence at temperatures sufficient to kill the living tissue and thus the starch content will be fixed.

Not only New Wood

Lyctus is a pest of new wood; that is, wood in its first few seasons after felling—whilst the starch content of the sapwood is at the required concentration. There does, however, seem quite an amount of evidence to suggest that timber very much older than this is attacked by *Lyctus*. Oak flooring strips in various forms are a case in point. The only explanation of this phenomenon appears to be that given above; that is, kiln-drying sometimes fixes the starch content.

During the last twenty years or so there has been shortage of some hardwoods (and indeed softwoods) as well as an increase in cost which has given rise to the tendency to cut down the period which hitherto elapsed between felling and fabrication of the wooden article. Thus not only has a large quantity of timber with starchy sapwood been made into furniture and other fabricated woodwork, but the actual proportion of sapwood used has been very much increased. There is no inherent disadvantage in this provided such wood has been suitably preserved.

The following home-grown hardwoods are attacked by *Lyctus* beetles:

Ash	*Fraxinus excelsior*	Evergreen Oak	*Quercus ilex*
Sweet Chestnut	*Castanea sativa*	Turkey Oak	*Quercus cerris*
Common Elm	*Ulmus procera*	Robinia	*Robinia pseudoacacia*
Dutch Elm	*Ulmus hollandica*	Sycamore	*Acer pseudoplatanus*
Wych Elm	*Ulmus glabra*	Walnut	*Juglans regia*
Hornbeam	*Carpinus betulus**	Cricket Bat	*Salix alba* var
Pedunculate		Willow	*caerulea*
Oak	*Quercus robur*	Crack Willow	*Salix fragilis*
Sessile Oak	*Quercus petraea*	White Willow	*Salix viridis*

* Little information

Note: NO SOFTWOODS ARE ATTACKED BY THESE INSECTS, although a single case has been recorded of an infestation in a piece of starchy softwood—a high starch content in softwood is unusual.

Recommendations have been given in which kiln-drying as a means of stamping out *Lyctus* (apart from the intrinsic merits of kiln-drying over air-seasoning) has been advised. A kiln-sterilization treatment is usually carried out at a higher temperature than for kiln-drying. The heating process kills the *Lyctus* larvae by heat. But, of course, this measure is transient only—the end-effect being exactly the opposite to that originally intended. Here is a list of imported timbers known to be susceptible to attack by *Lyctus* beetles:

Abura	*Mitragyna ciliata*
Afara	*Terminalia superba*
Agba	*Gossweilerodendron balsamiferum*
Black Bean	*Ccstanospermum australe*
Australian Blackwood	*Acacia melanoxylon*
White Bombay	*Terminalia procera*
Borneo Camphorwood (Kapur)	*Dryobalanops* spp.
African Celtis	*Celtis soyauxii*
White Chuglam	*Terminalia bialata*
Crabwood	*Carapa guianensis*
Dhup	*Canarium euphyllum*
Eng	*Dipterocarpus tuberculatus* and *D. obtusifolius*
Black Guarea	*Guarea thompsonii* probably susceptible
Spotted Gum	*Eucalyptus maculata*
Gurjun	*Dipterocarpus* spp.
Idigbo	*Terminalia ivorensis*
Iroko	*Chlorophora excelsa*
Ironbark	*Eucalyptus* spp.
Jelutong	*Dyera costulata*
Kokko	*Albizzia lebbek*
Indian Laurel	*Terminalia* spp.
African Mahogany	*Khaya ivorensis et al.*
Muninga	*Pterocarpus angolensis*
Tasmanian Oak	*Eucalyptus obliqua, E. regnans* and *E. gigantea*
Obeche	*Triplochiton scleroxylon*
Amboyna	*Pterocarpus indicus*
Indian Rosewood	*Dalbergia latifolia*
Sapele	*Entandrophragma cylindricum*
Seraya, Meranti	*Shorea* spp. and *Parashorea* spp. *et al.*
Teak	*Tectona grandis*
African Walnut	*Lovoa klaineana*

Some other timbers not included above are liable to attack by another group of beetles often included in the 'Powder-post' group, the family BOSTRICHIDAE. They are found in many of the above species.

5

LONGHORN BEETLES

The American Oak Longhorn—The Wasp Beetle—The Oak Longhorn—The House
Longhorn Beetle—Basket Beetles

The CERAMBYCIDAE is a family of beetles of widespread distribution
throughout the world. They are popularly known as the 'Longhorns' and
about 60 species regularly breed in the British Isles. In addition, many
species from abroad are regularly brought into this country in the immature stages in imported timber. They are mostly medium to large-sized
beetles and are characterized by having long antennae, in many cases
extraordinarily long. The larvae of practically all of them are wood-borers.
Some species are found in the sapwood of living trees, some in the bast
(the layer immediately beneath the corky layer), yet others are found only
in sound but dead sapwood, whilst others are found in wood already
infested with various species of fungus.

Quite a number of species of CERAMBYCIDAE occasionally make their
way out of furniture or structural timbers, either because the larva develops
naturally in dry wood or because, developing naturally in living sapwood,
it is nevertheless able to complete its life cycle in dry wood. In this case the
larval period is prolonged over many years. A large number of species of
Longhorn beetles occur in sufficient numbers in various parts of the world
to be of economic importance.

The American Oak Longhorn   *Eburia quadrigeminata*
Several species of Longhorn beetles are often quite common in the timber-
yard but the ability of some species, usually inhabiting only living trees,
to maintain themselves and, indeed, continue their development in the
dried and fabricated timber, sometimes leads to alarming results. *Eburia
quadrigeminata* Say. is an example of the latter type. Naturally attacking
North American oak, the standing tree, it has on many occasions walked
out of furniture some twenty years after manufacture (on one occasion
over thirty years). In this time an article of furniture has often become like
a valued friend, and when a large spidery beetle one day emerges from it
(from a flight hole a quarter of an inch in diameter), there is a feeling of
psychological horror in the household. Happily, this species is quite rare
in the United Kingdom. This beetle is nearly an inch in length and is
light-brown with a characteristic pattern of two pairs of white marks on
each wing case.

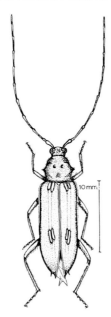

Fig. 32. American Oak Longhorn (*Eburia quadrigeminata*), a North American insect which occasionally emerges from oak furniture in the British Isles and is extremely long-lived as a larva. Several records of twenty years have been made and one larva lived thirty years!

The Wasp Beetle   *Clytus arietis*

This beetle not only has the black and yellow-banded appearance of a wasp, (Fig. 33) but by its agile, quick movements and its waving antennae it shows an extraordinary similarity to this insect. It is a common insect in the garden, where the larvae inhabit rustic ware and fencing, but is sometimes found emerging from furniture, which it will not re-infest. A number of closely related beetles, similar to *Clytus* in appearance, are known, from furniture manufactured from timber of foreign origin.

The closely related species *Anaglyptus mysticus* is another British species which can delay its development over a number of years when the wood into which it is boring is made into furniture. Records of up to thirteen years are known. It is somewhat like *Clytus arietis* but it has large mahogany-coloured patches at the fore-ends of the wing-cases.

The Oak Longhorn   *Phymatodes testaceus*

Perhaps the most common Longhorn beetle met with in timber-yards is the species *Phymatodes testaceus*, (Fig. 34). The fat inch-long legless grubs are commonly found in the bark and in the superficial sapwood. This again is a species which can complete its development in the dried and fabricated

D

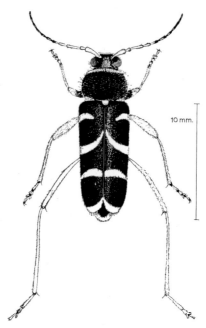

10 mm.

Fig. 33. The Wasp Beetle (*Clytus arietis* (Linnaeus)). Between one-half and three-quarters of an inch in length, this agile beetle often emerges from beech and oak furniture. The black and yellow bands give it a startling appearance.

timber. When it is found in buildings it is usually infesting new oak flooring. The beetle is extremely variable in colour from yellowish-brown to deep blue-black.

House Longhorn Beetle

Reference has already been made to the world-wide distribution of the wood-boring beetles in the family CERAMBYCIDAE. The number of species occurring naturally in the British Isles is upwards of sixty, and many more species are found in our timber-yards emerging from imported timber. One species, however, the House Longhorn Beetle, *Hylotrupes bajulus* (L.), shows a unique feature in its biology: alone of all the British Longhorns it is able not only to complete its development in dried timber, but to infest and re-infest old dried timber. It is probable that many House Longhorn infestations are due in the first place to timber of European origin imported with the young larvae already burrowing in the wood. A number of records have been made of House Longhorn larvae being found in packing cases and orange boxes. This may perhaps account for that part of its distribution which is scattered and discontinuous. With regard, however, to its local abundance in north-west Surrey, where, in some small towns and villages, a large proportion of the buildings has infested woodwork, it is certain that some combination of ecological features is favourable to its continued presence. It has recently been found that this circumscribed area is the warmest area in Britain. It is perhaps also signi- ficant that the areas in north-west Surrey where it occurs are sparse coni- ferous heathland with sandy soil, very similar to areas in Continental Europe which the insect inhabits.

House Longhorn attacks only softwoods and, at first, the attack is confined to the sapwood, but nevertheless heartwood also is often found to be attacked at a much later stage.

In appearance the adult beetle is very variable in size, from a quarter of an inch (males) to fully an inch in length (females). It is greyish-black to a very dark brownish-black in colour with two greyish transverse marks across the wing cases. Almost all the body is covered with greyish or yellowish-grey hairs, but the extent of this hairiness is variable. Some areas have the appearance of having the hairs rubbed off. A pair of such areas on the pronotum (the round part next to the head) have a remarkable resemblance to eyes (Fig. 35). The antennae are long but not unduly so.

Eggs of the beetle are yellowish to greyish-white in colour, elliptical with rather pointed ends, and are laid in cracks and crevices on the surface of the wood. The length of the egg stage is said to be about three weeks.

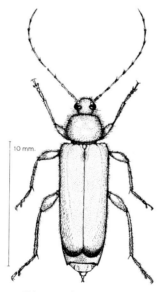

10 mm.

Fig. 34. *Phymatodes testaceus* (L). Two forms of this beetle are found; one fulvous brown, the other with the wing-cases a steely blue-black. It is upwards of half an inch in length and is quite common in oak, mainly confining itself to outer layers of sapwood. It often makes its way out of laid oak flooring, sometimes in considerable numbers, but does not re-infest the dried wood.

The larva is greyish-white and fleshy, with very pronounced segmentation (Fig. 36). The head is small with the greater part of it retracted into the body. The first three or so segments are large, the body then tapers, but the last two segments show an increase in size. The length of the full-grown larva varies from about three-quarters of an inch to an inch-and-a-half. When newly hatched, the larva bores straight into the wood. After a time the larval tunnel takes up an elliptical cross-section and large quantities of bore-dust are produced. The particles of bore-dust are not uniform. Some particles are very small, others are larger and have a rather granulated cylindrical appearance (Fig. 31), but the powder is so mixed with the air that it occupies a greater volume than the wood from which it is produced. This often is of importance in detecting an infestation because the packed powder causes the surface of the wood to give a blistered or rippled appearance.

When an infestation has been proceeding for some years, almost all the sapwood is converted to powder except for a thin veneer of surface skin. At this stage the surface skin often bursts along the grain (or where it was!).

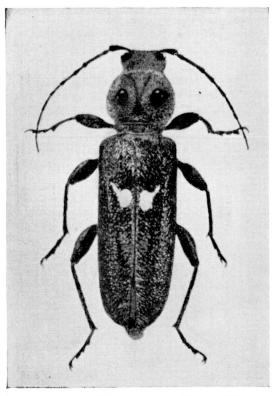

Fig. 35. House Longhorn Beetle, *Hylotrupes bajulus*. Note eye-like spots on the pronotum (the division next to the head). Crown copyright reserved.

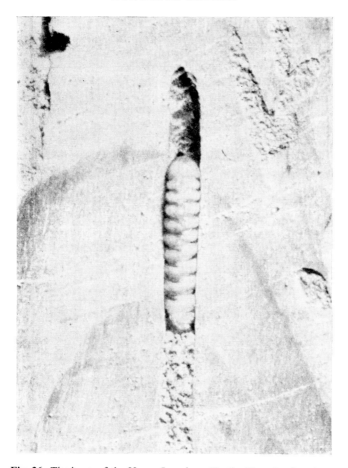

Fig. 36. The larva of the House Longhorn Beetle. Note the deep inter-segmental grooves and absence of legs. Crown copyright reserved.

It is remarkable how often House Longhorn infestations occur in the roof space of a house, and this is probably due to the peculiar micro-climate—very hot in summer and very cold in winter.

Often one is asked to identify an infestation in the laboratory from a small splinter of wood showing perhaps a small part of a larval gallery with no stages of the insect present. If the surface of the gallery is examined critically with the naked eye the bite marks of the larvae can be observed. The appearance is that of ripples in the sand left by the retreating tide. Several incorrect identifications have occurred in London where the galleries of the larva of the Wharf Borer (*Nacerdes melanura*) have been confused with House Longhorn. Wharf Borer galleries do not show these serial bite marks.

Pupation and Emergence

Pupation in nearly all the wood-boring Longhorn beetles follows approximately the same pattern. The larva eats its way close to the surface of the wood and then moves away again. It then blocks the gallery on the inner side, the particles of wood forming the plug being very coarse. The larva then changes to a pupa which is at first creamish-white but later becomes light brown. All the parts of the adult beetle can be made out in the pupa, but it is quiescent, not being able to move its legs. The abdomen, however, can be twisted about quite vigorously. The length of the pupal stage is thought to be about three weeks. When metamorphosis into the adult beetle has taken place the beetle bites its way out of the wood.

Fig. 37. Typical damage to the sapwood of softwood by the larvae of the House Longhorn Beetle.

Flight Hole

The flight hole is oval, the longer diameter varying from about an eighth to one-quarter of an inch. A relatively small number of larvae can cause a large amount of damage to softwood joists and rafters. Six larvae can convert to powder all the sapwood in a common rafter. Do not, therefore, expect to find a large number of holes in a House Longhorn infestation. A phenomenon exhibited by the larvae is the extraordinary straightness of the gallery when they find themselves in a new piece of timber. On one occasion when an infested joist was cut down one gallery was found to vary only half an inch from the straight in an eight-foot run.

Survey

It is unfortunate for the householder that House Longhorn beetle infestations seldom come to light before a great deal of damage has taken place. The roof space is seldom visited—perhaps only once a year or so to stow away junk. This means that the extent of the damage should always be surveyed by an expert, as some of the structural timbers may have to be replaced or reinforced. Notes on control measures are given in Chapter 8.

The Basket Beetles

Two rather small Longhorn Beetles—*Leptideela brevipennis* and *Gracilia minuta*—are sometimes found infesting baskets and other articles made of unbarked willow and hazel twigs. The small flight holes—about one-sixteenth of an inch across—are distinctly oval, and the larvae are very similar to typical longhorn larvae, except that they are very small.

6

SOME OTHER WOOD-BORING BEETLES

Their economic importance in furniture and structural timbers and how they may be
identified—The Wharf Borer, *Nacerdes melanura*
Ernobius mollis, Ptilinus pectinicornis and other ANOBIIDAE—The wood-boring weevils—
Buprestis aurulenta—BOSTRICHIDAE

The Wharf Borer   *Nacerdes melanura*

This beetle is thought to be native to the Great Lakes region of North
America and as it does much damage to piles and dock timber in this
region it is known in the United States, and now in Britain, as the 'Wharf
Borer'. It is in the beetle family OEDEMERIDAE, a family with only seven
British Species. In England it is common throughout the Thames estuary as
far inland as Teddington and is found in many estuaries on the south coast
as far west as Cornwall. The writer reared it from some rotten wood taken
from a houseboat moored at Worcester on the Severn, nearly forty years
ago. Until a few years ago it was thought to occur in Britain only in the
region of the sea coast, but a remarkable change seems to have occurred in
its habits since the last war. The large greyish-white larvae are found in
rotton wood in what can be called its typical habitats—at the water line in
estuaries. They occur not only in piling, wharfs and timber supporting the
river banks but also afloat! Quite a large number of Thames barges and
similar wooden-hulled ships sailing in the region of the Thames Estuary
support good colonies of Wharf Borer in those timbers which get washed
with fresh water. In its habitats, which cannot be said to be typical, it has
colonized sodden woodwork in every conceivable situation in the London
area. Buried wood—sometimes at considerable depth—has been found to
have numbers of living larvae in it years after the wood had been buried.
The adults can mate and lay eggs within the cavities in the wood gnawed
out by the larvae. Well-defined tunnels are not constructed by the larva,
nor does there appear to be much frass produced, but at intervals rather
coarse wood scrapings are found which have not passed through the body
of the larva. The larvae are also often found in poor property in floors under leaking lavatories. In sharp contrast to the buried woodwork Wharf
Borer larvae are found also in sodden timbers of bomb-damaged buildings,
sometimes at considerable heights above ground. Some of these infestations have been confused with House Longhorn by the inexpert, as there
is a similarity to old softwood well-tunnelled by the larvae of this latter
species. There are, however, no grounds for confusion, as the inside walls

of Wharf Borer galleries never display the ripple appearance of the bite-marks always shown by House Longhorn.

All wet timber attacked by Wharf Borer larvae is infested with fungal hyphae—either Dry Rot (*Merulius lacrymans*) or one of the other fungi causing rapid decay of wood where permanently wet conditions are con-cerned, such as *Coniophora cerebella*.

The larva of Wharf Borer (Fig. 39) cannot possibly be confused with any other of the wood-boring larvae we have described. It is often an inch in length and has a large yellow head. The six legs are also larger than hitherto seen in wood-boring larvae; indeed, in most cases larval legs are not well-developed. Paired patches of spinules occur on the top surface of the first five segments behind the head and on the underside of the sixth and seventh segments are pairs of stump-like false legs, also with spinules.

The adult beetle is variable in size from about a quarter to half an inch in length. It is a soft flimsy insect, golden-brown to reddish-brown, with the tips of the wing-cases black (*melanura* means 'black-tailed') and with long antennae. There are three raised longitudinal lines on each wing-case —a feature common to all beetles in the family OEDEMERIDAE. (It is sometimes confused with the typical brown form of the Oak Longhorn

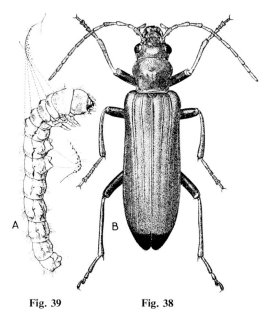

A   B

Fig. 39   **Fig. 38**

Fig. 38. The Wharf Borer, *Nacerdes melanura*. Light fulvous brown in colour with black-tipped wing-cases. Reproduced by permission of the Trustees of the British Museum.

Fig. 39. The Wharf Borer larva. The large head and patches of spines are distinctive features. Reproduced by permission of the Trustees of the British Museum.

Fig. 40. Piece of red deal from 'ceiling' of Thames
Barge heavily attacked by Wharf Borer.

Beetle, *Phymatodes testaceus* and with *Rhagonycha fulva*, a very common
and widely-distributed beetle with many local names.) In some parts of
London the insect has appeared in incredible numbers — tens of thousands
running over the pavements of the Strand, and yet their origin remained a
mystery. It has a strong flight.

Ernobius mollis

Another species in the family ANOBIIDAE is *Ernobius mollis*—a species
without a common name. It is rather like a large Common Furniture Beetle,
being up to a quarter of an inch in length (Fig. 41). When freshly emerged it
is covered with short golden hairs which give it a light shade of golden-
brown which darkens when some of the light-coloured hairs are lost. It
lacks the longitudinal rows of dark marks or 'punctures' on the wing-cases
which are present on *Anobium punctatum*. The wing-cases and other parts
are much less horny than in the Common Furniture Beetle and in other
species in the family ANOBIIDAE, and its trivial name *mollis* refers to this
characteristic, meaning 'soft'.

The eggs of *Ernobius mollis* are laid in summer in the bark of softwoods
only, and the larvae, which resemble those of Death-Watch, thereafter
tunnel in the bark and the outer ring of sapwood. Often this causes the bark

to fall off, exposing the tunnels packed with coarse frass; where the larva has been feeding in the bark the frass is the same dark colour as the bark; where it has been feeding on the sapwood it is light in colour. After the pupal stage has been passed the adult beetle bores out of the bark, leaving a round flight hole of approximately one-tenth of an inch in diameter. Hardwoods are never attacked. Softwoods are only attacked when the bark is still attached. This does occur perhaps more often in outbuildings and barns than in well-constructed dwellings, but even so, *Ernobius mollis* is known to occur in the latter. Softwood studdings, with bits of bark still attached, on which has been fixed plywood panelling, sometimes support infestations and the escaping beetles will bore through the panelling. It will be seen that little structural damage can be caused by this insect but sometimes anxiety is aroused when it is confused with Common Furniture Beetle. Any damage to timber would be superficial only. Control is easily effected by stripping off the bark where it occurs, when the infestation will rapidly die out.

Ptilinus pectinicornis

This is a small dark-brown beetle from one-eighth to nearly a quarter of an inch in length; it has a longer, more cylindrical body than Common Furniture Beetle, and has a curious, rather globular, prothorax, but the easiest distinguishing feature is the comb-like antennae of the male, (Fig. 42), a feature not so exaggerated in the female, where the antennae are merely toothed like a saw.

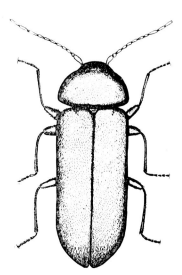

Fig. 41. *Ernobius mollis* is closely related to the Common Furniture Beetle but lacks the longitudinal rows of 'punctures' on the wing-cases.

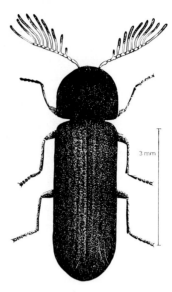

3 mm

Fig. 42. Male adult of *Ptilinus pectinicornis*. The un-
mistakable feathery antennae are clearly shown.

In *Ptilinus* as in *Ernobius mollis* the longitudinal rows of 'punctures'
on the wing-cases are absent. The larvae are found in relatively few timbers
compared with Common Furniture Beetle; beech, maple, ash and sycamore
being the most frequently attacked, but it is sometimes found in a mixed
infestation with *A. punctatum*. Out-of-doors, however, the range of timbers
that it will attack seems to be wider, but it is the hardwoods that are always
attacked—softwoods are but rarely bored by *Ptilinus*. The flight holes are
sometimes larger than those of *Anobium* but not so large as those of
Xestobium. The 'frass' or bore-dust is fine and silky to the touch, remini-
scent of *Lyctus*, again quite unlike that usually found in the ANOBIIDAE.
Indeed, recent work on the biology of *Ptilinus* indicates that it may be more
nearly related to the Powder-Post Beetles than to *Anobium punctatum*.

Besides Common Furniture Beetle, Death-Watch, *Ernobius mollis* and
Ptilinus pectinicornis, several other species in the family ANOBIIDAE have
wood-boring larvae, although in no case are they of such importance as the
former species.

Ochina ptinoides is a species with some resemblance to Death-Watch. It
flies well and may often be seen flying around old ivy-covered houses in
the late summer evenings. Its larvae bore into the old woody stems of ivy.
Three other species of ANOBIIDAE, *Grynobius excavatus*, *Anobium denti-
colle* and *Hedobia imperialis* are found boring into old dead hawthorn
stumps in the hedgerows. The last-named however, is occasionally found
boring in furniture.

The Wood-Boring Weevils

Two weevils, beetles in the family CURCULIONIDAE, are common wood-borers. They are often confused with Common Furniture Beetle; they are of importance, and the student of wood-boring insects should certainly be familiar with both of these insects and be able to differentiate between them. The two insects concerned are *Pentarthrum huttoni* Wollaston and *Euophryum confine* (Broun). Both beetles have a general weevil appearance with a long snout or 'rostrum' on which the antennae are situated, and a round cyclindrical body and short legs. In this they have a strong superficial resemblance to the grain weevil *Sitophilus granarius*, but both can be distinguished from the latter insect by having the wing-cases completely covering the abdomen, whereas in the grain weevil the last segments of the abdomen are uncovered.

The New Zealand Weevil *Euophryum confine* Broun. It was as recently as 1937 that this small beetle was first recorded as attacking wood in this country. In 1940 it was observed in a dead willow tree in Lea Valley Marshes, and in 1947 in Epping Forest. Later in 1947 it was reported by Dr. Hinton and Mr. McKenny-Hughes, of the British Museum, as attacking woodwork in London houses, and it was then pointed out that this weevil showed every sign of becoming a serious pest in houses.

These two entomologists were certainly right, as only a few years later it is now widespread, not only in London and the adjoining counties, but

Fig. 43. A piece of deal flooring attacked by *Euophryum confine*. Note irregular flight holes and that the wood has already been attacked by fungus.

as far as Leicestershire in the north. There seems to be no reason why this species should not extend over the British Isles; indeed, it may have done so already, as it has recently been reported in Dublin. It is, however, always found associated with wood in which there is evidence of decay, and the fungus usually responsible is *Coniophora cerebella* (Pers.), the cellar fungus. *Coniophora cerebella* can be killed by drying the wood, but the New Zealand weevil will often continue to infest wood even after drying if *Coniophora* has once been in it.

The adult beetle of *Euophryum confine* is small, from $\frac{1}{10}$th. to $\frac{1}{5}$th. inch in length; on the average being a little smaller than the related *Pentarthrum huttoni*. It is blackish-brown to reddish-brown in colour and the typical long snout or rostrum can be seen quite well with the naked eye. It walks with rather a 'loping' gait.

Unlike most of the other commonly-occuring wood-boring beetles the adult stage of this species burrows on its own account, and is often found quite deep in the wood. The adults can be found almost throughout the year, the reason being that there seem to be two overlapping life-cycles in the year.

The flight holes average about $\frac{1}{25}$th. inch in diameter—that is, definitely smaller than the flight holes of *Anobium punctatum* and only the smallest of the *Lyctus* flight holes approach it in size. *Euophryum* flight holes, however, are always ragged in outline, (Fig. 43), very distinct from *Anobium* and *Lyctus*, and often a sprinkling of powder occurs around the margin of the flight hole. Curiously enough this is a case of an insect of no importance in its country of origin becoming a pest of economic importance when introduced into another.

Pentarthrum huttoni. In general appearance this species is very like the foregoing, (Figs. 45–47). It is, however, more blackish and not nearly so reddish in colour as *Euophryum confine*. The larva and adult bore into timber which is damp and infected with one of the wood-rotting fungi. In addition, however, it is sometimes found in very dry timber, but in these cases it is always evident that some time in the past the timber has been wet and partially rotted by a fungus. Wherever such wood is found *Pentarthrum huttoni* may be present—it is commonly found in woodwork near leaking water-pipes in the vicinity of sinks, lavatories, bathrooms and similar situations, but it is amazing how often attacks by this beetle are found in breweries, wine vaults and beer cellars. They certainly seem to thrive in such places!

Very little is known about the duration of the different life-stages, but hardwoods and softwoods can be attacked, and damp plywood used as panelling against a wall is especially favoured. One curious characteristic is often observed concerning the appearance of the attacked wood. The softer layers of the wood are eaten entirely away first, leaving the harder rings untouched. This leaves the wood with a curious appearance, often like that of termite-attacked wood in the tropics.

Fig. 44. A piece of deal flooring attacked by *Pentarthrum huttoni*. Note 'worn' appearance of softer wood and irregularly shaped flight holes.

Buprestis aurulenta L.

The Golden Buprestid. Another wood-boring beetle which displays the phenomenon of delayed development is the species *Buprestis aurulenta* L. in the family BUPRESTIDAE—quite distinct from the Longhorn family (the CERAMBYCIDAE). This beetle is about an inch long and of a beautiful metallic green colour edged with copper. It is of North American origin, and when found in the United Kingdom it has usually emerged from Douglas Fir or other softwood used in house construction such as for a door or window sill. Normally, the larva is found in the living tree, but nevertheless is able to complete its development in the seasoned and fabricated timber, although such development is usually protracted over several years in such circumstances.

Bostrichid Beetles

Rather closely related to the *Lyctus* beetles is the family BOSTRICHIDAE, which contains a large number of mostly exotic wood-boring species. In general biology they resemble the *Lyctus* group in that they are only found in the sapwood of hardwoods rich in starch. Most species are larger than *Lyctus*, some being up to one inch in length, In shape they have something of the hooded prothorax of the Anobiid beetle, but contrasting with the latter in that the prothorax is curiously beset with spines. Bostrichid beetles are essentially pests of the freshly-felled tree, and under the conditions of

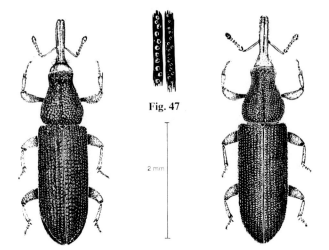

Fig. 47

2 mm

Fig. 45. *Euophryum confine.*   **Fig. 46.** *Pentarthrum huttoni.*

Fig. 47. Comparative size of sculpturing on the mesosternite of (left) *Euophryum confine* and (right) *Pentarthrum huttoni.*

the timber-yard they quickly die out. They do not re-infest the dried timber, but some species are able to continue their development in dry and fabricated woodwork and will sometimes emerge from furniture.

Nevertheless, Bostrichid beetles are a source of annoyance and loss to the timber importer, as often logs are received from abroad with all sapwood completely unusable on account of the large tunnels. (Damage to Obeche by *Bostrichoplites cornutus* and to European Oak by *Apate capucina* are cases in point.)

Fig. 48. The larva of *Euophryum confine* is entirely legless. Taken from wood rotted by *Coniophora cerebella.*

Identification of Bostrichid-damaged timber

Large tunnels, circular in cross-section in sapwood and running parallel with the grain, tightly packed with frass (except for the entrance tunnel in which parent beetles live and feed, which is clear of frass) are characteristic of damage by these beetles.

Control Measures. These are as for *Lyctus*, but are rarely required in Britain for fabricated articles such as furniture.

E

7

PINHOLE AND SHOTHOLE BORERS
(*AMBROSIA BEETLES*)

IDENTIFICATION OF WOOD-BORING
INSECTS BY THE DAMAGE THEY DO

Damage to timber by Pinhole and Shothole Borer, or Ambrosia beetles, as they are called, is one of the most serious problems confronting the widespread utilization of many tropical timbers. Quite a considerable proportion of these hardwoods (although the problem is by no means confined to the tropics nor to hardwoods) is damaged by these beetles. Unfortunately, a popular and widespread misconception exists in the public mind against the use of such damaged timber, as confusion with 'woodworm' invariably exists. The old beliefs (although fallacious) die hard— 'There is no "cure" for woodworm and every article showing the tell-tale flight holes must be burnt'. Thus timber showing pinhole damage is looked upon with the gravest suspicion. It is also surprising that many timber merchants have failed to grasp the significance of the problems involved, and little if anything has been attempted to overcome the difficulties. On the other hand much valuable scientific knowledge has been gained by the work carried out in recent years by the West African Timber Borer Research Unit. The biologies of a number of species have been worked out so that we know much more about the insects and their mode of life and, in addition, much experimental work on control has been undertaken. A relatively large number of beetles, mainly classified in the families PLATYPODIDAE and SCOLYTIDAE (the chief genera being *Platypus, Diapus, Trypodendron, Anisandrus* and *Xyleborus*), have larvae which infest the trunks and branches of, in most cases, freshly felled trees, but sometimes living trees. Until the last few years little detailed knowledge existed concerning the biology of these insects, and there are differences in detail among the various species.

As an example of the life history of this group, that of the Oak Pinhole Borer, a British Ambrosia beetle, is given below. The male beetle first flies on to the trunk of a freshly felled oak tree, probably attracted by the smell of fermenting sap. He then constructs a tunnel, and it is to this that the female is attracted. After mating, the female carries on with the boring. She lays small batches of eggs at intervals which hatch into slug-like larvae. The latter feed on a fungal growth occurring on wood fragments and tunnel lining.

Staining Fungi

A characteristic of the species of fungi which grow in close association with the tunnelling larvae is their property of staining the wood in the area where the fungus is alive. Purplish-brown is the usual colour of the stain, but it may vary from a relatively light hue to almost black. Sometimes the area of staining is confined to the walls of the tunnels, at other times the staining is more extensive—extending to fair-sized patches or flame-shaped areas, (Fig. 49). This fungus has been termed 'Ambrosia' and is the reason why the pinhole beetles are labelled 'Ambrosia beetles'. Another characteristic shown by both pinhole and shothole is the absence of bore-dust. The tunnels appear quite free of any 'powder', so different from *Lyctus*, which shows abundant fine powder. The *Lyctus* tunnels in addition are just the

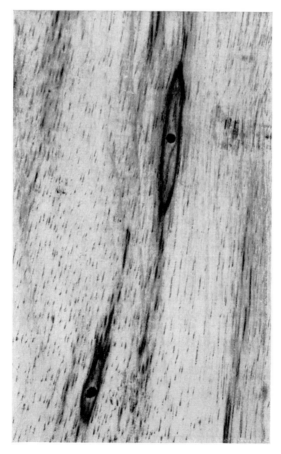

Fig. 49. Damage to timber by a Shothole Borer, characterized by the dark staining of the wood around the holes.

colour of the wood in which the tunnels are made. They are not stained in any way.

The most important feature of both Pinhole and Shothole Borers from the point of view of the timber importer is that when the moisture content of the timber drops to the extent that the Ambrosia fungi cannot thrive, the boring larvae die. Thus in the dry timber there are no living Pinhole or Shothole larvae and adult beetles are quite unable to infest such timber.

The distinction between Pinhole and Shothole is merely one of size. Pinhole Borer is the name given to those species whose galleries seldom exceed a diameter of one-sixteenth of an inch. The tunnels of the Shothole Borer, on the other hand, reach a diameter of one-eighth of an inch, which is approximately the size of lead shot. Incidentally, another important characteristic possessed both by Pinhole and Shothole is the straightness of the tunnels, which is quite distinct from the meanderings of either *Anobium* or *Lyctus* galleries.

Identification of wood-boring insects by the damage they do

When timber is being examined for evidence of insect damage, the adult insects are likely to be present only during the limited flight season of the species concerned. On the other hand, the larvae and pupae are inside the wood, only to be found and identified if the wood is carefully split up and thus destroyed. This is seldom possible in practice.

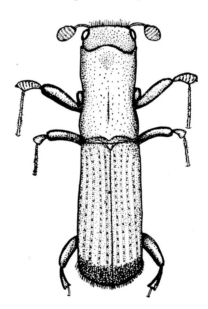

Fig. 50. *Platypus cylindrus.* The adult beetles of many Pinhole Borers are often of curious or bizarre shape. This species attacks oak, beech, chestnut and ash.

In the following table, therefore, the identification of the wood-boring insect is considered in its absence, the characters which are given being those of the damaged wood, the type of wood likely to be attacked, the nature of the tunnels and flight holes and the appearance of the faecal matter and rejected wood particles.

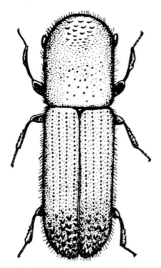

Fig. 51. *Xyleborus saxeseni.* This Pinhole Borer attacks a variety of hardwoods and soft-woods. The larvae, pupae and adults occur all together in brood-tunnels.

	TIMBER			LARVAL GALLERY			FRASS		FLIGHT HOLE	
Species	Sap or Heart	Fungal decay	Shape of Transverse Section	Straight or Meandering	Stained or Clean	Quantity	Description	Shape	Size	
Anobium punctatum	Wide range of softwoods and hardwoods	Mostly S but sometimes H	Not present or very slight	Circular	M	Clean	Moderate	Cigar shaped pellets. Gritty feel	Circular	About $\frac{1}{16}''$
Xestobium rufovillosum	Hardwoods oak chestnut, rarely softwoods if adjacent	S and H	Always present	Circular	M	Clean	Moderate	Bun shaped pellets	Circular	About $\frac{1}{8}''$
Ernobius mollis	Spruce, pine, larch, fir	Outer sapwood where bark adhering	Absent	Roughly circular sometimes Oval	M	Clean	Moderate	Bun shaped pellets present of two colours according to whether bark or sapwood has been eaten	Circular	Up to $\frac{1}{16}''$ in bark
Ptilinus pectinicornis	Beech, elder maple, oak, alder, plane, sycamore, elm, hornbeam, poplar	Sapwood	Absent	Circular	M	Clean	Moderate	Fine particles only. Silky	Circular	$\frac{1}{16}''$

Species	Timber	Wood		Tunnel	Direction	Walls	Frass quantity	Frass character	Exit hole shape	Exit hole size
Phymatodes testaceus	Usually oak but chestnut and ash recorded	Outer sap	Absent	Flattened oval and broad channels	Mostly with grain	Clean	Large quantities sufficient to loosen bark	Roughly cylindrical with one or two constrictions irregular ends	Oval	$\frac{3}{16}''$ to nearly $\frac{1}{4}''$
Clytus arietis	Usually beech and oak but Fruitwoods etc., not uncommon	Under bark until half grown then sap	Absent	Oval	Meandering	Clean	Moderate	Fine particles and loose aggregates	Flattened Oval	$\frac{3}{16}''$ long axis
Hylotrupes bajulus	Softwoods but recorded in some hardwoods such as oak, alder, poplar	Sap first but heart at late stage	Absent	Oval	Straight in early infestations later meandering	Clean 'ripple' teeth marks	Copious often bursting veneer of un-attacked surface wood	Fine particles (rejected) with well formed cylindrical faecal pellets	Oval	Up to $\frac{1}{4}''$ long axis but flight holes made by males, often very small
Buprestis aurulenta	Douglas fir and some other conifers	S and H	Absent	Flat Oval		Clean	Tightly packed in arc-like pattern		Oval	up to $\frac{5}{16}''$ long axis
Euophryum confine / Pentarthrum huttoni	Softwoods and hardwoods	S and H	Present often wet rot	Indistinctly round	M	—	Small	Fine in texture round granules	Narrowly Oval ragged margin	Slightly less than $\frac{1}{16}''$

	TIMBER			LARVAL GALLERY			FRASS		FLIGHT HOLE	
Species		Sap or Heart	Fungal Decay	Shape of Transverse Section	Straight or Meandering	Stained or Clean	Quantity	Description	Shape	Size
Lyctus sp.	Wide pored hardwoods Eg: oak, ash	Starch rich sap	Absent	Circular	M	Clean	Copious sometimes bursting un-damaged outer skin	Fine particles silky feel	Circular	About $\frac{1}{16}''$
Nacerdes melanura	Usually softwood occasionally oak. *Eucalyptus* recorded	S and H	Always present	Large irregular channels	Usually with grain	—	Moderate	Soft and wet with fibrous shavings	Ragged Oval	About $\frac{3}{16}''$ long axis
SIRICIDAE Wood Wasps	Pine, spruce, fir, larch	S and H	Present but not usually detectable	Circular	Curved	Clean	Gallery tightly packed	Coarse	Circular	Up to $\frac{5}{16}''$
SCOLYTIDAE PLATYPODIDAE Ambrosia Beetles	Many hardwood species more rarely in softwoods	S and H	Presence shown by staining	Circular	Very straight	Stained	Nil	—	Absent from sawn timber	—

Special Notes: The description of the wood damage caused by SIRICIDAE is equally applicable to the melandryid beetle *Serropalpus barbatus* which is common in Europe but absent from Britain.

8

CONTROL MEASURES

Anobium punctatum, extermination from furniture—Structural Timbers—Specification of wood preservative treatments—The Guarantee—Woodworm and the Law—Musical Instruments—Control of Death-Watch—Prevention and control of *Lyctus*, House Longhorn, Wood-boring Weevils, Wharf Borers

The subject of woodworm extermination and prevention is a large and complicated one, and is beset with many problems and side issues. It is realized that no simple answer to this question presents itself. In the first place, the term 'woodworm' includes a number of biologically distinct insects, and the differing life-cycles, types of timber attacked, differing ages of timber attacked and different parts of the wood tissue involved make identification of the species of the wood-boring insect a first essential. Knowledge of the biology of the insect concerned is therefore required.

The measures to be taken against any particular infesting insects must be closely related to this information. The identification and biology of those wood-boring insects which occur in the timber-yard, the cabinet-making factory, in furniture and structural timber in our homes—new and old—has now been discussed. A discussion follows on the measures to be taken for prevention and destruction of these pests.

Extermination of Furniture Beetle (*Anobium punctatum*) from Furniture

Introduction and Materials Used

Almost all the woodworm damage to furniture is caused by the two species Common Furniture Beetle (*Anobium punctatum*), and the Powder-Post Beetle (*Lyctus* spp.). The latter beetle infests 'new' furniture made from one or more of the susceptible wide-pored hardwoods, such as oak, chestnut, ash, abura, khaya and limba, and there are, of course, many others. Only the sapwood is attacked, and only when the starch content is above a certain minimum figure. This has already been explained in greater detail. 'Live' infestation of *Lyctus* will normally be found only in furniture manufactured within the last ten years.

Anobium will be found mainly in older furniture. An infestation may not be expected in the solid wood of oak furniture until some 60 years have elapsed from the date of manufacture. Birch and alder plywood will often be found to be attacked, however, within three or four years of its manufacture. This information means that when a furniture manufacturer takes delivery of some woodworm-infested articles back from the distributor, the

Fig. 52. A full kit of special tools, labour-saving devices and protective clothing used by a servicing team of woodworm exterminators.

insect is invariably *Lyctus*. Only one case to the contrary is known to the writer and this was where some sheets of birch plywood had been used which were obviously from a parcel which had been in stock a long time and located during a plywood shortage.

Some differences in procedure in dealing with woodworm will, of course, be usual according to whether this is carried out by a furniture manufacturer or whether it is being dealt with in the home.

'Home Remedies'

Almost every column written about woodworm in magazines mentions one or two 'home remedies'. Quite a common one is paraffin or turpentine which may or may not be applied to the holes with a fountain-pen filler. There is no doubt that if the sapwood of softwood is being treated, and provided also that some of the woodworm larvae are near the surface of the wood, paraffin would kill some of them, but such materials would by no means exterminate all of them and the effect would be temporary only. Both paraffin and turpentine are very volatile and, therefore, dry out within a comparatively short space of time. Shortly after such treatment a female beetle alighting on the wood may lay another batch of eggs, and as the paraffin or turpentine would have vaporized, re-infestation would commence at once.

Fumigation: This process consists of placing the infested article in a container or chamber which can be made gas- and pressure-tight and then introducing the fumigant by means of a valve. The fumigant is usually in the form of a gas, examples being hydrogen cyanide and methyl bromide. A specified time is allowed to elapse before the gas is drawn off or allowed to escape into the atmosphere. A further minimum period of ventilation before the articles are handled and subsequently delivered back to their owners is allowed as a safety precaution. Both gases given as examples are very poisonous and should be used only by skilled and experienced operators. Fumigation is a method of insect control often employed against, for example, stored-food insect pests, and its employment is sometimes recommended for use against woodworm. It must be remembered, however, that fumigation by itself acts in a temporary manner only. As soon as the fumigant is dispersed (within a few hours of the treatment) there is nothing to prevent re-infestation, so that fumigation is usually employed in conjunction with the use of long-acting oil-soluble woodworm-killing fluids or special insecticidal polishes. This combination of treatment is most useful on items which for various reasons may not be treated with oily fluids or where upholstered furniture or large quantities of furniture are to be treated. Obviously, by the nature of the fumigation process it must be carried out by firms of specialists licensed for the purpose.

Heat Treatment: In leaflet No. 13. The Forest Products Research Laboratory suggested the use of this process where woodworm occurred in polished furniture and it is stated that temperatures of 130°F., and relative humidities of 80% do not appreciably affect surfaces finished with cellulose paint, french polish or turpentine varnish.

In addition, several types of short-wave radiation have, from time to time, been suggested as a means of tackling the woodworm problem but there seems to have been little commercial interest taken in these methods, as indeed in heat treatment. The reason for this appears to be the comparative ease of application and the extraordinarily high degree of efficiency and long-lasting powers of the oil-soluble woodworm-killing fluids, whereas heat treatment and short-wave radiation would have a temporary effect only.

Discussion

An advantage held by the fluids with a non-volatile base, as against those with a volatile base, is that as soon as the latter dry out or lose their solvent to the atmosphere, penetration into the wood obviously ceases, leaving the insecticidal materials at the level of penetration reached when this occurs.

On the other hand, those fluids containing an appreciable percentage of non-volatile mineral oils exhibit the phenomenon of 'creeping' or continuous penetration of the wood. This has been demonstrated as occurring even after several years have elapsed from the time of initial superficial brush treatment. However, the deeper the insecticidal chemicals penetrate, the more extended or diluted they become, and only reputable brands with appropriate chemical loadings to allow for this factor should be used.

Extermination from Furniture   *Practical Notes*

When a piece of furniture is found to be attacked by woodworm, treatment for its extermination should be carried out without loss of time, It will be appreciated that the occurrence of flight holes will signify that at least one generation of *Anobium* or *Lyctus* has completed its life-cycle and it would be almost certain that one or more overlapping generations of larvae would still be present in the wood.

The view put forward, from time to time, that insecticidal treatment against woodworm should be undertaken only at the time of emergence of adult beetles from the wood, is mistaken. The semi-permanent effect of the proprietary oil-soluble woodworm fluids makes their effect independent of season of application. The simple treatment outlined below should be given to any furniture immediately it is suspected of harbouring woodworm, or if it is proposed to prevent possible woodworm attack in the future.

Overall Brush Treatment: Remove all dust and dirt; then, in the case of upholstered articles it is desirable to remove any leather or textile to prevent staining*. It is then best, after removing any drawers or shelves, to turn the article upside down and thus obtain more or less, the Furniture Beetle's-eye view. Apply one or more liberal coats of the proprietary woodworm fluid which has been chosen, according to the manufacturer's instructions which will usually be found printed on the tin, and if there is any special coverage, such as one gallon to so many square feet, make sure that this amount at least is applied. Work the fluid into all the cracks, crevices and rough surfaces with the brush. The bottom surfaces of legs should not be forgotten, as perhaps, in furniture, this is the most common part for the beetle to attack first. It would be the first part to be reached by a female beetle walking over the floor in search of an egg-laying site and, of course, helped by the fact that the bottom surfaces of the legs are almost invariably left in a very rough condition.

An important distinction between the various proprietary fluids must now be understood. Some materials are not suitable for treatment of polished surfaces, whilst others do not harm such surfaces. Fluids coming within the former category will have this fact brought to notice on the manufacturer's 'directions for use'. If the fluid which is being used belongs to the latter group it should be liberally brushed over all the polished surfaces, after the unpolished surfaces have been treated. The drawers and shelves which have temporarily been removed, should be treated in the same way.

Injection: Almost all furniture is made of hardwoods which in the main resist the penetration of fluids to a greater degree than softwoods. It is,

* If this is impracticable then ensure that no fluid is applied nearer than one inch to any such material, or arrange to have the piece fumigated by a specialist firm.

therefore, always advisable to inject the fluid through some of the flight holes and thereby also obtain an immediate kill of larvae which may be boring deep in the wood, (Fig. 53). This is, of course, in addition to the overall brush coats. A range of injectors has been manufactured during the last twelve years or so since British Patent No. 651262. Several different types exist, but perhaps the most successful model uses a plastic container (constructed of polythene) as the bellows, (Fig. 54). It is possible to remove the down-tube in some injectors, thus allowing them to be used upside down.

The mode of action is as follows: select some of the larger flight holes and insert the nozzle of the injector into one of them, pressing the rubber washer against the surface of the wood; the subsequent pumping of the fluid into the wood will ensure full pressure, forcing the fluid along the intercommunicating galleries, often deep into the wood. The rubber washer, then, is of great importance and is the reason why fountain-pen fillers, oil cans, hypodermic syringes, etc., lacking this part are not nearly so efficient. Spraying devices introduced into flight holes are sometimes used, but with little success because the fluid being sprayed is deposited on the walls of the gallery at its beginning only and the air merely blows out bore-dust. The aerosol type injectors are however an exception to this. The positive action of injectors is often demonstrated in dramatic fashion. Fluid will often emerge from a flight hole some distance, as much as a foot away, from where it is being pumped in. This is, of course, because of the way in which the tunnels intercommunicate within the wood. It follows that the worse the infestation, that is, the more intercommunicating tunnels there are, the easier it is to introduce the woodworm-killing fluids right

Fig. 53. Using an injector against Furniture Beetle.

inside the wood if a good injector is used. Care should be taken not to exert too great a pressure which might cause splashing or deflection backwards into the eyes.

Filling: If any fluid remains unabsorbed on the surface of the wood after a few days, it should be removed with a dry cloth, and it is probably a good idea to fill in the flight holes, if they are not too numerous, with a filling material. For light woods molten beeswax can be pressed into the hole with a knife; this is, of course, the good old standard way, but with darker furniture it is a good plan to fill in with one of the plasters such as 'Alabastine' or 'Polyfilla'. These are white, but have the property of absorbing colour and can be tinted to give the exact match (an 'Oxo' cube is excellent for the purpose) and be polished afterwards.

The filling of flight holes serves the useful purpose of indicating fresh ones, should they occur.

If a chair or a pew has been treated it is often advisable to seal the woodworm-killing fluid with a wood sealer sold for the purpose by the manufacturers. This will prevent clothes from spoiling if the chair is to be used immediately.

Fig. 54. Perhaps the simplest and at the same time the most successful type of injector for use against woodworm in furniture. The plastic container allows the operator to obtain an excellent pressure. Note the all-important nozzle washer which enables a tight closure to be made without marking the surface of the polished furniture.

Period of Immunity

It is often asked how long furniture or woodwork generally, treated with one of the modern oil-soluble wood preservatives, will be kept free from woodworm attack. The writer, of course, is not able to speak for all types of fluid, so that it is difficult to give a general answer, but having a great deal of experience with one particular type of fluid, it can be said that in this case, at a very conservative estimate, the period is from twenty to thirty years; indeed there is strong evidence to suggest that the period may be much longer.

Wetting Action of Fluid

The fluids which have been mentioned above are almost all light-brown to colourless and at least have no pigment or dye in their formulation. In order to be effective, however, the surface of the wood must be wetted by the fluid, and as the light colour of many woods is due to air in the surface fibres, the wetting action will have the effect of toning the wood a slightly darker colour. As the fluid is absorbed, however, by the drier wood underneath, the original colour will return. This means that it sometimes happens, in imperfectly polished wood, that some part of the surface is not wetted, and this gives a streakiness which may persist for some months. The extermination of the woodworm is, however, immeasurably more important than a temporary colour blemish.

Due care should be taken to prevent splashing the fluid on plaster work.

Extermination of Furniture Beetle from Structural Timber

By far the most important part of the woodworm problem in the United Kingdom concerns the widespread occurrence of *Anobium punctatum* in the structural timbers of buildings. As previously mentioned, this almost certainly concerns more than half the total number of dwellings, numerically something like 8,000,000, besides an incalculable number of farm buildings, sheds, industrial properties and other buildings. Beside this, the problem as concerns furniture pales almost into insignificance.

Before carrying out any work against woodworm in the structural timbers of a building, an accurate survey must be undertaken to ascertain the nature and extent of the infestation and its distribution. Access should be made into the roof space, the 'cupboard under the stairs', basements and cellars. Indeed, all woodwork must be thoroughly examined and reported on.

The surveying operation alone often entails much physical discomfort and painstaking work, and whilst some householders may be capable of carrying it out competently the great majority consider that this task and the subsequent control work must be the responsibility of a woodworm-servicing company.

Over the past few years the *in situ* treatment of building timbers against infestations by *Anobium punctatum* by servicing companies has become a

Fig. 55. Cleaning joists in a confined space. It is important to remove all dust and débris before treatment.

highly specialized industry, and although the householder by this means is able to place the responsibility for carrying out the work on other shoulders it would be very wise for him to make himself familiar with the general principles of the work.

The Decision

As a householder, the first stage in woodworm control is to make the decision to have it carried out. The question often arises, when making this decision, whether the woodworm control treatment is best carried out during a particular season, such as just before emergence of the adults. Such a season for treatment, say in early June, would be valid only for those beetles emerging in that season, whereas there will remain in the wood larvae which will not be emerging till the following year and the year after that. In fact, the long-lasting properties of woodworm-killing fluids make their application independent of season. The best time, therefore, to have treatment carried out is as soon as possible after the decision has been made, as is mutually convenient to the householder and the treatment firm.

There are other aspects concerning the woodworm control decision which should be mentioned here. Why is the decision made? Why do between two and three householders out of four, when confronted with the decision, decide to get woodworm control under way as soon as possible? There are several reasons. Perhaps the most important relates to the large capital investment represented by the purchase of a house. This is usually the most important capital investment ever made by the average man, so that when he realizes that some part of the structure is progressively deteriorating he feels compelled to do something about it.

In addition to this reason for the decision, however, over the past few years there has occurred a very strong feeling that insect pests of any description in the home should not be tolerated. Complete freedom from insect pests is, indeed, one of the criteria of high living standards.

In a high proportion of cases the presence of woodworm only comes to light during a change of house-ownership. It commonly occurs that a surveyor employed by a prospective purchaser establishes the presence of the pest, for example, in the roof void, and then recommends his client to make an offer much less than the price being asked. On the other hand, the vendor who is already aware of the presence of woodworm in his house often, wisely, gets the woodworm trouble cleared out of the way before putting his house on the market. Wisely, because the sum offered by a prospective purchaser whose surveyor has found a woodworm infestation is invariably considerably less than the asking price less the cost of wood-worm treatment. Moreover, the presence of woodworm in the home, shown by the groups of little holes appearing in well-loved furniture or in the structure of the home, the little brown beetles and the piles of frass,

Fig. 56. A thorough treatment of all wooden surfaces with long-lasting insecticidal fluid in each 'building unit' will control Common Furniture Beetle with virtual certainty.

F

Fig. 57. Wickerwork damaged by Common Furniture Beetle. Old baskets thrown into the loft or roof void are often reservoirs of infestation for many years.

cause worry and anxiety, and in extreme cases when other worries or anxieties are present as well, can give rise to mental ill-health of considerable severity. All those who have experience of working in woodworm control in direct contact with the public are able to give many instances of this.

The Survey

The first essential part of a woodworm control operation is the survey. This must be conducted by a thoroughly trained and experienced timber infestation surveyor who not only can identify with accuracy all the different species of wood-boring beetles and wood-rotting fungi likely to be found in buildings, but who can estimate the distribution and extent of the damage likely to have been caused. The degree of attack by any of the insect and fungal decay factors must be assessed, and the best means of control specified. The timber infestation surveyor, unless specially instructed to do so, will not normally inspect outbuildings, gates and fences outside the building, nor furniture inside. An underfloor inspection is very necessary as, although floorboards often show little or no woodworm attack on the top surface, the underside and the floor joists may often show considerable infestation. The timber infestation surveyor, therefore, will require to lift a floorboard or two and use his torch in order to make an underfloor inspection. Much can be done by the householder to help the timber infestation surveyor. Fitted carpets and other floor coverings can be loosened, and where possible rolled back. Bulky stored articles should be moved away from timbers, especially in the roof, and more especially from around the trap door. It is equally important to remove stored articles from the vulnerable 'cupboard under the stairs'. It is remarkable how frequently this space shows the first signs of Furniture Beetle attack. In the case of large buildings adequate ladders should be available for inspection of otherwise inaccessible woodwork. It is surely obvious that it is in the best interests of the householder to co-operate with the timber infestation surveyor and to grant him every facility in carrying out his task. The surveyor divides the building into its separate building units; e.g. the roof voids, the various floors, the staircase and the joinery and these are separately examined and reported upon, because a rather different specification is applicable to each.

The Specification

We give below the specification used by Rentokil Limited as an example of a specification of wood preservation treatment for wood-boring insect attack, divided into the separate building units as described above. The timber infestation surveyor would, of course, also include in his report any observations and specifications concerning attack by wood-rotting fungi— but fungal decay is outside the scope of this present book.

SPECIFICATION OF WOOD PRESERVATIVE TREATMENTS

1. Wood-boring Insects. The basic treatments have a common application to the following species of wood-boring insects, viz.: Common Furniture Beetle *Anobium punctatum*, Death-Watch Beetle *Xestobium rufovillosum*, Powder-Post Beetle *Lyctus* spp., and House Longhorn Beetle *Hylotrupes bajulus*. The treatments are detailed below and will, where appropriate, be specified in our report as complete treatments. Any variation from these treatments will be specified in our report. House Longhorn treatment is specially dealt with in paragraph 2 below.

a. Roof Voids. Lift floorboards as necessary in the roof void, clean down all exposed roof timbers with a vacuum cleaner where practicable and necessary and apply the insecticidal fluid to all such exposed surfaces of the rafters, purlins, roof boarding, struts, tie beams, ceiling joists, etc., and all surfaces of any roof floorboards before relaying.

This treatment is referred to as FULL ROOF VOID TREATMENT.

b. Ground Floors and Upper Floors. Lift the floorboards as necessary and clean down surfaces. Clean all accessible surfaces of the joists and floor framing timbers and apply the insecticidal fluid to all exposed surfaces of the floorboards, joists and floor framing timbers. Relay existing floorboards, renewing as necessary with pre-treated new flooring.

This treatment is referred to as FULL FLOOR TREATMENT.

c. Joinery Timbers. Timbers which are painted or varnished are less susceptible to insect infestations than sawn structural timbers or flooring. Where evidence of infestation is observed in joinery timbers, the visibly infested timbers require to be treated by one of the three following alternative methods which will be specified in our report.

This treatment is referred to as FULL JOINERY TREATMENT. Numbers 1, 2, and 3 respectively.

Full Joinery Treatment No. 1. Carefully remove the visibly infested skirting, picture rail, architrave, door lining, panelling or other similar joinery timber as particularly described in the report and apply the insecticidal fluid to the unpainted reverse side and end grains as also to all timber fixing plugs, grounds, etc. Replace and fix in position as before, leaving any necessary touching up and making good of decorations to match existing for client's decorator.

Full Joinery Treatment No. 2. (Where F.J.T. No. 1 is not economically justifiable.) Using a bradawl form holes each of depth two-thirds the thickness of the infested timber to supplement existing flight holes as means of pressure injection of the insecticidal fluid, the holes to be placed so far as possible in the least conspicuous positions, and (together with flight holes) at 3–4 inch centres. After injection extend holes to full depth and give low pressure trickle treatment through each hole to rear surface of the infested timber. Any necessary stopping of holes and making good of decorations to be left to client's decorator.

Full Joinery Treatment No. 3. (Where redecoration of infested timbers is already proposed.) Strip off existing paint from visibly infested timbers, using stripper to maker's directions or blow lamp as necessary and apply the insecticidal fluid to all stripped surfaces, also injecting into flight holes at 3–4 inch intervals.

d. Staircases. Apply the insecticidal fluid to all exposed timber surfaces, ensuring particularly as far as possible the treatment of all accessible unpainted rear or undersurface of timbers painted or varnished on one side. Painted or varnished timbers not accessible for treatment on their reverse sides, e.g. wall strings, to receive F.J.T. No. 2 or No. 3 as may be specified in our report which will also indicate any need for the removal and, where necessary, the subsequent renewal of plaster or other soffites.

This treatment is referred to as FULL STAIRCASE TREATMENT which term will be used in conjunction, where necessary, with F.J.T. Nos. 2 or 3.

e. Half-timbered Houses. Carefully wire brush all accessible surfaces of such external hardwood timbers as are not sealed by varnish or paint films, and which are recommended for treatment, and apply the insecticidal fluid, also injecting into flight holes at 3–4 inch intervals. Any painted or varnished and infested timbers requiring treatment are, to extents to be specified in our report, to be treated as for F.J.T. No. 3.

This treatment is referred to as HALF-TIMBERED HOUSES TREATMENT (EXT.) which term will be used in conjunction, where necessary, with F.J.T. No. 3.

2. House Longhorn Beetle *Hylotrupes bajulus.* In the special case of Longhorn infestations the basic treatments specified above will be supplemented as follows: after cleaning, carefully cut away all heavily tunnelled sapwood down to sound timber, and examine all apparently sound accessible timber surfaces with a strong pointed instrument to determine the extent of sub-surface destruction of timber. Carefully collect, remove, and burn all frass. Reinforce and/or renew affected timbers as necessary to restore structural strength. Apply one liberal coat of the insecticidal fluid to all such de-frassed areas, before applying the fluids provided for in the basic treatments specified above. The appropriate types of insecticidal fluids will be specified in the report.

SPECIAL NOTE

It will be seen from this basic specification that a considerable amount of what can be called carpentry work is entailed—such as the lifting and sub-sequent relaying of floorboards. So in comparing estimates submitted by two or more servicing companies, the appropriate specification also should be compared, as some servicing companies undertake no carpentry work; leaving this to the householder's builder. With other companies the estimate is inclusive of all such work. Like should be carefully compared with like. It has always been the opinion of the author that woodworm control is a specialist business and full responsibility for it should be borne by the treatment firm, not hived off on to a building company.

In any building where there is an attack of woodworm in some part of it, the *ideal* remedy would be for treatment to be carried out *throughout all the timbers*. Consideration of the life-cycle has previously shown that the external evidence of an attack occurs only at the *completion* of the first life-cycle. Economics, however, seldom allow this complete procedure to be adopted. The most reasonable arrangement seems to be the consideration of the building in its individual timber units just as it was surveyed, and the recommendation of treatment only in those units (e.g. the roof void) where the infestation is seen to be evident, widespread or locally abundant. At the present time it is not possible to state in precise terms what constitutes an attack of woodworm. Obviously we ought to have some standards. On the visible evidence of one single flight hole it would be difficult to sub-stantiate this as an attack, but on the other hand as little as a dozen flight holes distributed throughout joists of a floor could be held to consitute one. Obviously the timber infestation surveyor must be allowed a consider-able measure of discretion when making his recommendations, when a number of factors such as the age of buildings, the heartwood/sapwood ratio of timbers, etc., would be taken into consideration.

When the report and estimate have been passed to the householder it is a good plan for the timber infestation surveyor to explain the reasons for his recommendations.

Guaranteed for Twenty Years

Although it is unfortunate that it is not possible to observe a woodworm attack until the end of the first life-cycle, when the first adults emerge from the wood through the flight holes they have made, infestations of wood-boring beetle larvae do show certain advantageous features as far as their control is concerned when compared with other forms of insect control. Perhaps the most important concerns the fact that during almost all the life-cycle of wood-boring beetles they are virtually imprisoned in the wood. This simplifies, and indeed makes certain, their complete control. This is because it is possible to apply insecticides to wooden surfaces with great precision as regards siting and lethal dose. Even when larvae are boring deep in the wooden structure, so that they are beyond the effect of the penetrating fluid, they must inevitably return towards the surface in order to construct the pupal chamber and finally emerge as adult beetles. If, therefore, the timber surface has been treated at an accurate dosage rate with a carefully calculated lethal zone of woodworm-killing insecticide the wood-boring larva must just as inevitably succumb.

Another important advantage possessed by woodworm-killing fluids of the oil-soluble type really concerns the properties of wood itself. This is the ability of wood, and more especially the sapwood, to absorb oil solutions. It is sometimes possible to demonstrate quite remarkable penetration powers of such solutions in sapwood. To a somewhat lesser extent this is true of water solutions also, but in this case the wetter the wood the better the penetrating power of the water solution. This phenomenon has two effects, the first of which is the obvious one of allowing the woodworm-killing fluid to reach a high proportion of the burrowing larvae. The second concerns the lasting properties of several insecticidal substances. Certain of such substances lose their insecticidal properties after a few months when applied as a superficial film on a non-absorbent base, but when absorbed into wood through the medium of penetrating oil solvents, their woodworm-killing properties last for many years.

Thus it is that the long-term guarantee concerning suitably treated wood is not only a practical proposition based on well-known scientific phenomena, but has come to be generally expected as part of a woodworm control service.

The Form of the Guarantee

It is obvious that such a guarantee must be given only in respect of the treated timbers, which is the reason why a specification detailing the timber to be treated is an integral part of the woodworm-control contract. In the author's experience it is a very rare event for re-treatment of timber still showing emergence of wood-boring beetles to be necessary.

In the certificate of guarantee issued by the originators of long-term guarantees there is an important clause of particular interest to those selling their home; this is to the effect that in this event, the servicing company is prepared to assign the benefit of the remaining period of the guarantee to the purchaser of the property and will recognize the purchaser as the person entitled to the benefit of the guarantee.

Woodworm and the Law

The occurrence of heavy and widespread damage to the structural timbers of houses and other premises and to furniture and the subsequent fear of it by a large proportion of the public has naturally led to a number of disputes concerning it. There are the landlord and tenant disputes—the allegation that woodworm in the tenant's furniture has spread to landlord's floors, or, it might well be, the other way round. Really the only comment that can be made is that such spread is impossible to prove with certainty, however *probable* it might appear in some circumstances. The recommendation to guard against such eventuality if one has either infested premises to lease or infested furniture to put in them is to follow the advice given with regard to application of suitable woodworm-killing fluids.

The selling of premises and furniture infested with woodworm gives rise to another set of problems which may end in litigation. It is not proposed to go into a great deal of detail in this regard, but some sound advice emerges from the writer's experience as an expert witness.

Never guarantee or otherwise give a warranty that any article of wood, be it furniture, structural timber or musical instrument, is free from woodworm *merely because flight holes are absent*. The reader will no doubt by now have realized that it is possible for an article of wood to have a number of woodworm larvae tunnelling in it, yet flight holes would not appear until such larvae had completed their development and emerged as beetles. This has already been decided in the courts against the seller of a piano giving such a warranty.

If you are selling a house or a valuable piece of furniture which you know to have woodworm do not try to cover the flight holes or otherwise mislead the intending purchaser in any way concerning its occurrence. You may, you think, complete the sale satisfactorily, but the purchaser may be one of the growing public becoming aware of woodworm, and perhaps some little time after the completion of the transaction the purchaser may be able to prove that you, as the vendor, made out that the house or furniture was other than what it was. Judgements have already been given on such cases which, although rather complicated, indicate clearly that you, as vendor, would not fare too happily in Court.

Before offering such a house or furniture for sale you would be well advised to consult a specialist firm able to survey and estimate for the extermination of the woodworm. Armed with this figure you would then allow such an amount to the intending purchaser, to be deducted from the purchase price.

If you yourself carry out any treatment in your house against woodworm, or if a specialist firm carried out such work on your behalf, it is important to *keep all the documents*. Attach the receipted bills for woodworm-killing fluids, etc., to the documents of your house. At the very least they may be well worth their face value in the event of your or your heirs selling the property at a later date.

Musical Instruments and Their Treatment

With the increase in Common Furniture Beetle it only required the action of the laws of probability for such articles as old violins, which had for two or three hundred years not received a visit from an egg-laying Furniture Beetle, suddenly to become the objects for their attention. This is not to say that none of these old instruments ever showed signs of woodworm damage in the past.

The owners of old violins and other valuable and historic wooden objects are naturally anxious to preserve their property—the Stradivarius violin probably represents wood in its most valuable form—but one is always asked: 'will this insecticide treatment affect the tone in any way?' The answer cannot be definite, but it is believed that the tone is scarcely affected, if at all. Certainly no one who has carried out this treatment has ever informed the writer that deterioration in tone has occurred.

The inside of a violin makes an excellent rearing site for Furniture Beetles. I should imagine the female enters the belly of the instrument through an 'S'-hole and there would find many admirable egg sites. The application of a fluid to the outside of a fiddle does not present many difficulties, but there will have to be some ingenuity displayed—with a camel-hair brush tied on to a piece of wire—before all surfaces of the inside have had a brush treatment. But take one important precaution—do not apply the fluid to the pegs, or they will slip.

Pianos constitute another problem. A very large number of pianos came into this country from Germany at the turn of the century. They had beech frames, and quite a high proportion of them are now infested with Furniture Beetle. The essential process for extermination remains the same—a coat of proprietary oil-soluble woodworm fluid applied by brush to all surfaces and injected here and there where there are flight holes—say every four inches or so.

A piano, again, presents some problems concerning the application of fluid to the inside, mainly on account of weight—but a few splashes of fluid on to the wire strings will cause no harm. Incidentally, the same fluid will prevent rust and keep moths away from the felts.

Sometimes one is asked whether creosote—on account of its cheapness— may be used in woodworm control.

Creosote—a dark brown waste product of coal-tar distillation—has been widely used in the past for woodworm extermination. Generally it has not been successful in preventing emergence of Common Furniture Beetle, nor does it give to timber much more than two or three years' immunity from attack when applied by brush as an external coat.

For creosote to be a successful wood preservative the timber needs to be pressure-impregnated. Creosote-impregnated timber can only be recommended for use where there is no human contact, as with railway sleepers and telegraph poles, on account of the bleeding of the timber and the disagreeable smell.

Detection by Amplifying Instruments

Instruments have from time to time been devised to amplify the sounds made by the jaws of wood-boring larvae. In this way it is hoped that a means of estimating the effect of insecticide or fumigation might be found. Most instruments appear to suffer from a common defect. All noises are amplified, and in a town it seems impossible to use the instrument for detection of wood-boring larvae in large timbers. But small pieces of wood heavily insulated from extraneous noises would be suitable subjects for tests. It must be remembered, however, that wood-boring larvae spend by no means all the time boring, so that a wood-boring larva may be present although it cannot be heard.

Control of Death-Watch Beetle

In general this follows the same principles as those laid down for Furniture Beetle, *Anobium punctatum*, but because of the association of Death-Watch with fungally-decayed wood emphasis must be given to the eradication of the environmental conditions giving rise to fungal attack. The application of fungicides is sometimes valuable. Ventilation of the wood must be increased materially, and any sources of excess moisture removed. The insecticidal fluids must be applied at heavy dosage rates and injected into the wood, where possible through old flight holes. Death-Watch attack often takes place in cavities within large dimensioned timbers, and these must be reached by injection if suspected.

Injection can, of course, only take place where there are flight holes, but a method of introducing the fluid right inside the wood, often carried out with success, is to bore a series of holes as near to the top of the beam as possible, and sloping downwards. These holes are then filled with fluid which is replenished from time to time. Funnels made to fit tightly in these holes by means of corks allow more fluid to be introduced than would otherwise be possible.

Death-Watch is another wood-boring beetle which will, on occasion, bore through such materials as lead in order to emerge from the wood.

One of the most amazing things about Death-Watch is being able to detect it. It is amazing how often a severe attack of Death-Watch can occur in a church roof without it being observed. Sometimes at the time of emergence the adult beetles are present in thousands; at other times, where there is an equally severe infestation (shown by fresh crops of flight holes) the adult beetles are by no means so apparent. The presence of bats feeding on the beetles has been held to account for this. Again, where there is a severe Death-Watch attack, the spread of the attack to softwood should be expected, as in these circumstances it is not uncommon. These softwoods also show fungal infection. This is, however, quite a different phenomenon from the apparent infestation of deal boards lying on top of infested oak joists. The Death-Watch flight holes only represent lines of escape from the joists. The deal boards in this case are usually sound and free from fungal infection. In treating such cases, as well as preservative treatment, the joists should be insulated from damp by means of slates.

There is a general impression that Death-Watch extermination is exceptionally difficult. The reason for this impression appears to be inadequate rather than ineffectual treatment. Some timbers are treated that are known to be infested, and perhaps the following year a number of Death-Watch beetles are found on the floor and the observation is made that the treatment has done no good. A check should be made to find out where the beetles have come from. In the experience of the writer these beetles have often emerged from timber which has not been treated. This brings us back to the golden rule for Furniture Beetle extermination, which should, as far as economically possible, be applied to Death-Watch—treat the whole of the job.

Prevention and Control of 'Lyctus'

When hardwood timber is felled and the moisture content commences to drop, a stage of drying is reached when the female *Lyctus* beetle is attracted to the felled timber and lays her eggs. She will also lay eggs in sapwood of well-dried timber provided a minimum starch concentration is present. Common Furniture Beetle (*Anobium punctatum*) is not usually found in freshly-felled timber, so by far the most important pest in the timber-yard, or indeed anywhere that stocks of hardwoods are maintained, is *Lyctus*.

Three official leaflets have been issued by the Forest Products Research Laboratory on the subject of *Lyctus* beetle (the notes which accompany these titles are, of course, based on the author's personal opinion).

Leaflet No. 3. *Lyctus Powder-Post Beetle*, contains a brief description of the biology and discussion of control measures. The present high price of hardwoods, however, make some of the recommendations impracticable— such as elimination of sapwood—and emphasis is laid on some control measures such as kiln sterilization which are of only a temporary nature, and if carried out on a wide-scale would only succeed in shifting the infestation from the timber-yard to the furniture owner.

Lyctus infestations must be confined to the timber-yard by methods which are now commercially practicable. The references to what might be called 'timber-yard hygiene' are, however, excellent.

Timber Sterilization

Leaflet No. 13. *Kiln Sterilization of* Lyctus-*infested Wood*, gives a schedule for heat sterilization of timber in order to kill *Lyctus* already alive in it for all thicknesses of timber up to three inches. It is interesting to note that for 3-inch timber with a kiln temperature of 115°F (46°C) and relative humidity of 60%, the total period of exposure is $49\frac{1}{2}$ hours. Although *Lyctus* is the subject of this leaflet it is implied that other wood-boring insects may be killed by heat sterilization. Readers may not be familiar with the writer's opinion that heat sterilization is a factor prolonging *Lyctus*-infestations, making the attack more severe when it does occur than if other methods of extermination are used.

The statement made in this leaflet that heat sterilization does not render timber immune from subsequent re-infestation is of the greatest importance and it is this fact that, in the writer's opinion, renders it undesirable. There are, of course, commercially practicable methods of achieving this by insecticidal means.

Lindane Spray Treatment

Leaflet No. 43. *Prevention of* Lyctus *attack in Sawn Hardwoods by use of Contact Insecticides*, gives the method of applying an emulsion of Lindane in water to timber in stick in order to prevent *Lyctus* attack. This method is a prevention only and is of little use where *Lyctus* larvae are already present in the timber. The Lindane is purchased from the manufacturer as a concentrate of 20–25% Lindane in an aromatic solvent such as naptha, xylene or similar material, together with emulsifying agents. This concentrate is made up by adding sufficient water to give at least $\frac{1}{2}$% Lindane in the finished spray. The emulsion made is a white milk-like fluid which should be stirred vigorously and frequently whilst it is being used. It should only be made up in sufficient quantities for spraying that day otherwise the emulsion 'creams' —a high proportion of the Lindane coming to the top of the fluid. It is highly recommended that all timber is liberally sprayed with this emulsion before *Lyctus* is expected to emerge to commence egg-laying. Spraying should be finished by the end of March—at latest by the middle of April. If a hot summer is experienced and second brood of emerging *Lyctus* is feared then a second spraying should be carried out during the month of July. It is not of course, necessary to devise means of spraying in between the planks, as it is only the sapwood in which the female *Lyctus* beetle is interested, but all the other sapwood edges should be thoroughly treated. Difficulties are, however, experienced where timber is piled or cross-piled. Long thin lances have then to be used in order to get to sapwood edges in the centre of a stack; even so 100% coverage cannot be obtained in every case.

The timber coverage by the resultant $\frac{1}{2}$% Lindane emulsion will vary, of course, with the area of cut surface of sapwood so that the thinner the timber the more fluid will be required per unit volume of timber. Tests carried out under commercial conditions have, however, given a figure of 120–200 cubic feet of timber per gallon as a general guide.

If the timber is re-sawn or machined in any way so that fresh surfaces are exposed then these surfaces must be re-treated. No discoloration of the wood occurs nor is the wood harmed in any way by the chemical constituents of the emulsion. There is no doubt that the widespread application of this treatment, which originated in the laboratories of the Forest Products Research Laboratory, has had a very appreciable effect in decreasing *Lyctus* in the United Kingdom.

Unfortunately, not all manufacturers and timber-yard managers have sufficient foresight to take part in this preservative work. What usually happens is that insecticidal treatment is only put into motion when *Lyctus* is already found to be present in the timber-yard in great quantities. Much damage has then already been done.

Timber should be stacked and piled in such a way that all sapwood edges can be reached by commercial spraying equipment. Sticking only ¼-inch in depth means that it is impracticable to spray the sapwood of such timber.

Insecticidal Treatment

Treating Existing Infestations in the Timber-yard: The conclusion reached widely by many entomologists studying insecticidal composition is that the most effective base for applying insecticides to insects is an oil. Oily materials have the effect of penetrating the waxy cuticle of the insect and allowing the insecticide to reach the nervous system. This is especially so in the case of what are known as contact insecticides. It is not necessary for the material to be taken into the insect's mouth; the contact of the insect with the insecticide is all that is required.

Where a live infestation of *Lyctus* is seen to be present in the timber-yard the most effective and long-sighted method of dealing with it is by using one of the specially formulated oily-base insecticides which have been formulated for dealing with wood-boring insects. Alternatively—but more expensively—the timber can be re-sawn in order to remove the sapwood.

The important point is not only to kill the *Lyctus* in the timber but to so render the timber unacceptable for *Lyctus* re-infestation. This is of supreme importance. These special *Lyctus*-killing fluids are not diluted in any way, as the effective base is an oily one and the oil is necessary, not only for carrying the insecticidal ingredients, but also to obtain penetration of the wood; certain non-volatile mineral oils are used for this purpose. This material is already in use by a number of manufacturers who treat the fabricated articles in the white, either dipping, spraying or brushing on the fluid. Such insecticidal fluids can be coloured in order to carry out two purposes at once, the staining of the wood in addition to the *Lyctus* prevention.

Such fluids are practically colourless, but, of course, the wetting action of the fluid on the surface of the wood gives it a darker tone (water would, of course, give the same effect). Two days after treatment, cellulose, french polish, varnish or paint may be applied and will give a hard finish.

'Lyctus' in Hardwood Floors and Joinery

New oak and other hardwood floors and joinery are sometimes attacked by *Lyctus*. The attacks will, of course, only be present in the sapwood sections. Often the particular pieces attacked are so badly powdered that some replacement is necessary.

Any wax polish, varnish or french polish should be removed by solvent or by sanding machine and the wood treated with woodworm-killing fluids. A thorough treatment is necessary as *Lyctus* is sometimes difficult to eradicate by treatment from the top surface only, but care should be taken to avoid getting the fluid between the cracks in a wood-block floor. The bitumen adhesive used can be softened by heavy application.

Control of House Longhorn

In the specification for the control of House Longhorn Beetle which has already been given, it will be seen that any sapwood tunnelled to such an extent that it can no longer bear any strain should be hacked away. Such sapwood would only soak up large quantities of insecticidal fluid and serve no useful purpose. When all the powdered sapwood has been removed the whole area should be thoroughly cleaned, all dust and débris being removed, preferably by vacuum cleaner. All the woodwork—rafters, ceiling joists, roofing battens, etc.—should be given two or three liberal coats of special insecticidal fluid. This should be carried out by brush or coarse-nozzle spray. Fumigation is sometimes suggested for House Longhorn control, but it must be borne in mind that the effect (like heat treatment) is temporary only. As soon as the fumigant is dispersed (or as soon as the woodwork returns to room temperature), the wood is liable to re-infestation, Insecticidal fluids with long-lasting insect-killing properties must, therefore, be used. As these fluids are satisfactory without fumigation, the significance of fumigation and heat-treatment is seriously diminished.

A very important point concerning House Longhorn infestation is its obvious ability to spread from house to house, as its compact distribution in north-west Surrey clearly shows. In the interests of all concerned the right of search by competent officers under the jurisdiction of the Local Authority is very wise.

Control of Wood-Boring Weevils

Make a thorough examiniation in the region of the premises where the beetles are being found in order to locate all the infested wood. Remove it —it is rotten anyway—and sterilize surrounding woodwork with one of the proprietory oil-soluble fungicides. One containing pentachlorophenol is recommended. Then make sure that sources of excess moisture have been removed and the ventilation improved. Cut the replacement wood to size, liberally apply the fungicidal fluid to all the surfaces, and finally fit in and fix.

Control of Wharf Borers

This species will not occur where there is good building maintenance; that is, where timber is free from fungal decay. The renovation and cleaning up of bomb-damaged property and sites inevitably caused a reduction in the numbers of this insect. Half-buried barges and hulks in the Thames estuary are, however, a reservoir of infestation. No timber, unless adequately protected with a fungicide, should be used for wharfing or piling or used in building construction where damp conditions prevail. Where an infestation has occurred and the strength of the timber has been seriously impaired, remove the cause of damp and cut out infested timber, then apply two or three liberal coats of an oil-soluble fungicide (one containing pentachlorophenol is recommended), over all the surrounding woodwork and the replacement timbers, if the timber is not to be cut out, dry it out thoroughly and apply liberal coats of the fungicide to all surfaces of the wood. In the case of sailing barges and other wooden-hulled craft a water-soluble fungicide may be added to the bilge.

9

WOOD WASPS AND MARINE BORERS

SIRICIDAE—Teredo—Gribble

Wood Wasps SIRICIDAE

Wood Wasps are fairly large insects classified in the insect order HYMEN-OPTERA—the ants, bees and wasps. Insects in this order have four wings, the hinder pair smaller than the fore pair, and the wings of each side interlock by means of a row of hooklets. The more primitive HYMEN-OPTERA, the saw-flies, Wood Wasps, and the honey bee, are vegetable feeders, but the vast majority of species in this very large order are carnivorous, being mostly predacious or parasitic on other insects. Although a number of Wood Wasp species have been imported in timber from various parts of the world, the species most usually found in Britain are as follows:

Urocerus gigas. The female is a large black-and-yellow insect, very much like a hornet in general appearance, and although it possesses a large sting-like projection (the egg-laying apparatus) at the tail end, in fact it is harmless. The abdomen of the male is brownish-orange and does not possess the sharply-pointed egg-laying tube. Although usually large in size, diminutive examples also occur, so that size alone is unreliable for identification.

Sirex juvencus, Sirex noctilio and *Sirex cyaneus* are three species which, being very similar in general appearance, require a trained entomologist to identify them. The interested reader is referred to more specialist works. The females of these species are steely-blue in colour and the egg-laying tube is prominent. The abdomen of the male is brownish-orange with some black at each end. During warm sunny days the female Wood Wasps bore holes in the bark of pine, larch, fir and spruce trees which are lacking in vigour. The boring is carried out with the egg-laying tube which is exceedingly tough and furnished with cutting teeth. A few eggs are laid at the bottom of each shaft. A white secretion lubricates the walls of the egg tubes to facilitate the passage of the long thin eggs. By means of this secretion the wood is also inoculated with a fungus carried by the female in special glands. The fungus is the species *Stereum sanguinolentum* which causes a white-pocket rot in many species of coniferous trees. The *Sirex* Wood Wasps cannot exist without it, as the hyphae of the fungus grow rapidly preceding the boring of the *Sirex* larva and play some part in its nutrition. The larvae of *Sirex* Wood Wasps are whitish, cylindrical, and

95

have three pairs of very blunt projections underneath the thorax which serve as legs. The larvae, however, are easily identified by the terminal spine at the end of the abdomen, which is of characteristic shape and projects upwards. The length of the life-cycle out of doors varies between two and three years, but indoors this may be prolonged for several years. Softwood timber infested with a few larvae of *Sirex* Wood Wasps is not uncommonly used in building construction, and eventually the adult wasps emerge, which, because of their rather fearsome aspect, usually cause some alarm. They do not however, re-infest dry wood inside buildings. A point of interest in the biology of *Sirex* Wood Wasps concerns their parasitism by two remarkable insects, also in the HYMENOPTERA. The female of *Rhyssa persuasoria* is equal in size to *Sirex* but more slender, and possesses a very long toothed egg-laying tube. She endeavours to locate the *Sirex* larva in the trunk and then bores down to it and deposits an egg on it. *Rhyssa* passes through its life-cycle in one year. The second parasite is *Ibalia leucospoides*, and in this case the female searches for and eventually finds the egg-shaft made by the *Sirex* female. She then inserts the egg-laying tube down it and lays an egg in every *Sirex* egg she can reach. The *Ibalia* larva passes the whole of its life, which is about two years, inside the *Sirex* larva.

Wood Wasps emerging from timber in buildings are not a serious problem. They are, however, of importance to manufacturers of goods being sent to Australia on account of official regulations which forbid the importation of softwood from countries where these insects exist unless the timber has been fumigated or otherwise treated to terminate possible infestations. As packing cases are made of softwood the problem can be a large one in many industries.

Fig. 58. Piece of Central American Mahogany taken from furniture, showing early attack by *Teredo*. The small apertures made by the *Teredo* larvae can be seen on the right; these have been planed off on the left. The borings had been covered by green baize.

Marine Borers

Although, strictly speaking, Marine Borers are not woodworm, yet it was thought advisable to give some account of them here for students of timber preservation.

When timber is immersed in sea water, it is not long before certain organisms effect a lodgment on it, and, if the timber has not been protected mechnically or chemically, boring commences which, if allowed to go un-checked, will destroy it. Whereas, on land, in temperate countries, by far the most important wood-boring animals are the immature stages of beetles and in tropical countries Termites (both insect groups), in sea water insects are absent. Two entirely different orders of animals, however, take their place as far as wood-boring is concerned—the CRUSTACEA and the MOLLUSCA.

Fig. 59. Shipworm—*Teredo*. A wood-boring member of the MOLLUSCA. In proportion to the size of the body the shells are much reduced. It occupies the whole length of the burrow.

Teredo

The *Teredo* is a bivalve mollusc (as is an oyster or mussel), and therefore is not strictly a worm as is implied by its common name, shipworm. The eggs are released haphazardly into the water in very large numbers and from them ciliated larvae are produced which are able to propel them-selves through the water. After a time two microscopic shells are produced together with a 'foot'. Its size at this stage is one-hundredth of an inch in length and when the larva ultimately reaches a wooden surface it attaches itself in position by a thread known as a byssus which is secreted from a gland in the 'foot'. The tiny shell valves are then brought into action by being

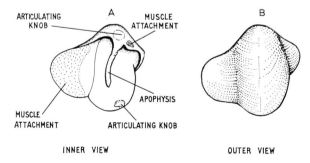

Fig. 60. The 'shells' or boring valves of *Teredo*. Enlarged.

G

Fig. 61 Piece of softwood heavily 'gribbled'.

made to rock backwards and forwards, and by this means the larva quickly burrows beneath the surface of the wood and never returns to the open water again. It continues to bore by the rocking action of the valves, the latter being pressed against the wood by the 'foot'. As the burrow lengthens, so the body of the *Teredo* elongates so that the length of the *Teredo* always equals the length of the burrow. As it grows, the *Teredo* secretes a chalky lining to its burrow which is characteristic. Near the minute opening are two small siphon tubes, through one of which (inhalent) the *Teredo* draws in sea water which is filtered of its small organisms before being passed out through the exhalent siphon. A pair of 'pallets' are able to block up the opening after the siphons have been withdrawn.

When fully grown the *Teredo* can reach a length of four feet and a diameter of one inch, but the usual size is much less than this. Size mainly depends on the number of *Teredo* present in the wood. They never share a burrow—each being quite distinct although the chalky lining may be the only walls at some points between their bodies. Thus, if only a few *Teredo* are present, they will grow to a large size, but if many are present, they will not grow so large.

It is quite common for hardwood logs imported into this country to show superficial *Teredo* borings. Storage and collection of the logs often takes place in harbours infested by *Teredo*, and during the month or so of floating in the sea the *Teredo* larvae settle on them and start to bore. *Teredo* is able to survive fairly short periods of drying-out of its timber—ten days to a fortnight or so—a factor which has to be reckoned with in dry-docking of wooden-hulled ships. It can also keep alive for a considerable time if the salinity of the water is reduced somewhat. The average salinity of the sea is 35 parts per 1,000 and down to salinity of 10 parts per 1,000 it can burrow and reproduce normally; if, however, the salinity is reduced to 5 parts per 1,000 it can only just exist, and below this salt concentration it will die within a few weeks. This information is important to engineers of harbour works in estuaries where the salinity fluctuates with the volume of fresh water brought down by the river. Temperature also affects *Teredo*. The slightly higher temperature of the sea water in a harbour where there is an effluent from a generating station is said to cause increased *Teredo* activity. *Teredo* activity is much more pronounced in tropical waters.

The Gribble   *Limnoria lignorum*

This wood-destroyer is a crustacean and is classified in the group ISOPODA, to which the common woodlouse belongs, and indeed, it has the same number of walking legs—seven pairs—as this latter animal. The Gribble is never much more than one-fifth of an inch long and its tunnel is usually not longer than two inches. The burrowing Gribble keeps close to the surface of the wood, entering at a small angle, and here and there small lateral galleries connect with the outer surfaces—these are the respiratory pits or 'manholes', (Fig. 62). By means of these the Gribble ensures a plentiful supply of fresh oxygenated sea water. Each gallery houses a male and a female Gribble, but it is the female which does the burrowing, the male usually being found some little way behind her. When disturbed the male is easily extracted from the burrow, but not so the female. She flattens herself against the end of the burrow and with the sharp claws on her legs is well able to resist removal. Usually no more than twelve larvae are produced at a time, and when these leave the parental brood-pouch they start burrowing side tunnels from that of the parents. Later, when fully-grown, they may leave the burrow and, swimming through water, find other timber to infest.

When examining Gribble-infested wood another small animal is sure to be extracted from the Gribble burrow. This is another crustacean, a member of the group AMPHIPODA, to which the sand-hopper and the fresh-water shrimp belong. Its scientific name is *Chelura terebrans*, (Fig. 64) and

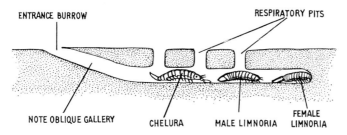

Fig. 62. Gribble (*Limnoria lignorum*) and *Chelura terebrans* shown diagrammatically in their burrows.

G*

it is thought that some symbiotic relationship exists between it and *Limnoria,* i.e. that they enjoy mutual benefits from living in association with each other.

Gribble-infested wood is decayed in stages. The infestation is always confined to the outer skin of wood, but when this has been completely riddled with the Gribble burrows, wave action washes it off so that a fresh surface of wood is presented to the Gribble to destroy. Gribble requires water of high salinity, so that it is not found in estuarine waters. The appearance of Gribble-infested wood is quite characteristic and is very easy to identify.

Fig. 63. The Gribble (*Limnoria lignorum*) top view. This is an isopod crustacean in the same group to which the Woodlouse belongs.

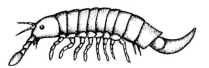

Fig. 64. *Chelura terebrans*—side view. An amphipod crustacean usually found in the Gribble burrows.

10

TERMITES (WHITE ANTS)

If the whole world is considered, then without a doubt by far the most important wood-destroying animals are termites. About 1,800 different species occur in countless numbers in all the warm countries of the world. Happily, in this instance, neither Great Britain nor indeed Northern Europe comes into this category. Two species, however, *Kalotermes flavicollis* and *Reticulitermes lucifugus*, are found commonly in Southern Europe, and another species infests a small area of the city of Hamburg, It is in view of the fact that all articles, constructed wholly or in part of wood, intended for use in tropical or sub-tropical countries, must be rendered termite-proof that this short account is given here. In addition to wood, many other materials including certain plastic products, are often damaged by termites.

Termites are often given the name of 'White Ants'. Actually they are not ants and, in fact, are very dissimilar in structure and thus, classification. It is, however, very probable that in the habits of the termites living in very large communities the layman has seen a similarity to the social organization of the true Ants (FORMICIDAE).

The order of insects to which the Termites belong is known as the ISOPTERA. They do not undergo a metamorphosis during development, as do beetles and butterflies, where the grub or larval stage is very unlike the adult insect. The young termites when they hatch from eggs are very like the adult in form, and any differences that do occur are assumed more or less gradually during the development period, as they shed their successive skins.

Termites are very interesting biologically on account of the complex social organization in which they live. In the case of what are usually known as subterranean termites, the adults exist in a number (usually three or more) of what are termed 'castes'. These are the forms in which they exist. King and Queen termites are usually produced from winged forms. These forms possess two pairs of almost identical wings which are of service for a few minutes only during the sexual swarming. When, however, they drop to the ground and pair, the wings break off close to the base, (Fig. 65) as thereafter they live in a special chamber deep in the colony, from which they do not emerge again. The abdomen of the Queen enlarges to immense proportions—she produces many tens of thousands of eggs. From her eggs are produced worker and soldier termites which never attain

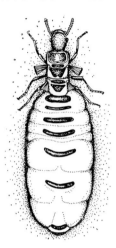

Fig. 65. Queen termite. Note stumps of wings where they have broken off.

sexual maturity. These are produced in very large numbers and spend their whole lives tending the communities in which they live. The workers forage for food, enlarge the nest and keep it clean (which involves eating the bodies of their own kind), or in the case of the wood-boring species carry out all tunnelling. The wood they eat is digested by PROTOZOA (one-celled animals) in the stomach and a salivary material produced from the digested wood is fed to the King and Queen, whose digestive organs are modified to take only this food.

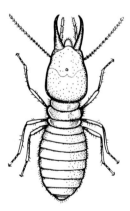

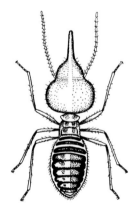

Fig. 66. Soldier termite (Mandibulate type); they defend the colony but do no other work.

Fig. 67. Soldier termite (nasute type); they defend the colony but do no other work.

The sole duty of the soldier termite is the defence of the colony. On being disturbed the soldiers take up threatening attitudes in the front line and will seize in their mandibles any marauding insects such as true ants whilst the workers immediately stow young brood and eggs in the deeper recesses of the colony. The soldiers of some termite species are able to eject a stream of fluid which repels the attackers. These are known as nasutes, (Fig. 67) and they possess a large beak-like projection at the front of the head which secretes a strong-smelling and sticky fluid from a gland at the tip.

From time to time another caste—subsidiary reproductives—are produced; these are to augment egg production. They differ from the Queen in never having possessed wings. When climatic conditions are propitious the colony swarms. Large quantities of 'alate' or winged reproductives are produced which, after a brief spell of flying, mate, and then shed their wings before searching out a crevice in a piece of wood in order to start a new colony.

Termite Classification, with notes on main genera

The termites are classified into six families. The MASTOTERMITIDAE contains only a single Australian species found in tree stumps underground. The community often consists of over a million individuals, but there is doubt as to whether these are true workers. Fossil forms of species in this family have been found in England, and the family is important as showing the relationship between the termites and the cockroaches.

Members of the second family, the KALOTERMITIDAE, are known as the dry-wood termites, on account of their inhabiting dry wood which may have no direct connection with the ground. Usually the communities are small, and there is no worker caste, the individuals mostly being juveniles. Soldiers, however, are present which are characterized by the possession of large heads and strong mandibles. The faecal pellets are also characteristic, being dry and seed-like, and from time to time a quantity is ejected from the wood through a hole made for the purpose which is usually closed up again afterwards. Well-known species are *Kalotermes flavicollis* from Southern Europe, *Cryptotermes brevis* the West Indian dry-wood termite, and *Cryptotermes dudleyi*, from East Africa, Central America and South-East Asia.

The TERMOPSIDAE are known as the damp-wood termites and are usually found in fallen logs and stumps in conditions of high humidity. Like the KALOTERMITIDAE, the colony size is small and true workers are not present. One species, *Zootermopsis angusticollis*, gets transported over long distances, sometimes reaching this country in logs of Douglas Fir and Californian Redwood from the Pacific coast of North America.

The HODOTERMITIDAE, the fourth family, are called the Harvester Termites.

The fifth family, the RHINOTERMITIDAE, are the subterranean termites. They are usually small and are important wood destroyers; *Reticulitermes lucifugus* is the destructive European species, and *flavipes* in the same genus destroys much timber in buildings in the U.S.A.

The great majority of termite species, however, are classified in the family TERMITIDAE. In the sub-family MACROTERMITINAE wood is chewed into 'combs' on which certain species of fungi are grown and devoured. The *Microcerotermes* are found throughout the tropics, and *Amitermes* are particularly able to live in desert regions.

The *Odontotermes* are of considerable economic importance. The *Nasutitermes* are remarkable for the pear-shaped head of the soldiers, and some species build arboreal nests.

Subterranean Termites

The main colony or nest of this type of termite is located underground, often very deep, or in otherwise inaccessible places. From such a situation the workers make long galleries through the earth until some wood is located when the tunnel is continued into the timber. Only rarely is evidence given of the presence of the termites until the wood is so badly eaten away by the tunnelling that it collapses or breaks. Termites never work in the light, nor, indeed, in the open; they require a very close control of humidity, and if any open space has to be crossed it is accomplished by the construction of a tunnel or tube, which can be seen, made of earth and saliva. In

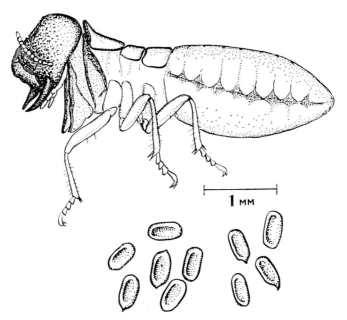

Fig. 68. A dry-wood termite *Cryptotermes brevis* with characteristic fæcal pellets.

this way the subterranean termites can bridge over concrete building foundations and commence an attack on sound woodwork on the ground-floor.

Dry-Wood Termites

These differ in habit from the subterranean types in not requiring any connection with the earth. The colony is able to support itself in lofty spaces high above ground, roof timbers being common situations. The infestation is often only made apparent by piles of the seed-like faecal pellets dropped on to the floor.

Control

The termite problem in the warm countries is such a large one that man's effort to safeguard timber in all its uses has resulted in a vast quantity of literature involving all types of recommendations. It is, however, amazing what little concern is shown with regard to this problem in many areas. Timber and labour is cheap and infested woodwork is completely replaced rather than simple preservative treatment being carried out.

For the reader in Britain we give below brief precautions that can be taken in the termite-infested territories, but what will be more the concern of the reader will be the precautions to take regarding timber-fabricated articles—including packing cases—manufactured in Britain but destined for the tropical and sub-tropical countries.

In areas where termites abound no timber in whatever form should be allowed to remain in contact with the ground. Keep the ground clear of all rubbish, fallen twigs, etc. Before building commences the soil-poisoning operation should be undertaken. This consists of saturating thoroughly with long-lasting modern insecticides the soil in the area where the building is to be constructed. In the construction of houses and other buildings allow no timber to be used within eighteen inches of the ground. Use bricks or concrete for foundations. Constant vigilance must be maintained in order to see that no shelter-tubes are constructed over the foundations.

All timber used in building construction should have two or three applications of an oil-soluble wood preservative manufactured for the purpose or should be pressure impregnated with a water-borne preservative. There are a number of proprietary brands. Dipping with oil-based preservatives is often carried out and can also be recommended.

For the British manufacturer producing articles wholly or partly of wood for use in tropical or sub-tropical countries the specification should include complete treatment of wooden parts with oil-soluble or water-borne wood preservatives. All wooden packing cases should be similarly treated.

Treatment should be carried out when the wooden article is completely fabricated; that is, when no fresh wood surfaces will be produced. Furniture destined for abroad, as well as motor truck chassis and scientific instruments, should be treated in this manner.

11

INSECTS COMMONLY CONFUSED WITH WOODWORM

Almost every insect which has at one time or another made its way into a house accidentally or otherwise must have been confused with woodworm —at least, that is what the writer believes after receiving specimens for identification from many tens of thousands of householders during the last twenty years or so.

These insects fall into two main categories—those insects which are infesting some materials in the home (other than wood) such as textiles or stored food, and another group consisting of wanderers from the garden or elsewhere—their presence in the house being accidental or fortuitous. The following brief description includes insects which, from experience, the writer knows to be most commonly but erroneously thought to be wood-boring pests. No attempt is made here to produce an entomological text-book, but the illustrations and short notes should make identification of the household insects quite simple in most cases.

The Biscuit Beetle

Stegobium paniceum, sometimes known by an obsolete generic name *Sitodrepa*. This beetle is in the same family—ANOBIIDAE—as Common Furniture Beetle and Death-Watch and it does have a strong resemblance to the former (Fig. 69), and, indeed, it is often mistaken for it by house-holders. It is not a wood-borer but is a common pest of many stored foods, especially farinaceous ones. In warm kitchens or cupboards it often emerges at odd times during the year. It sometimes subsists on crumbs which have collected between the floorboards of old houses.

It can be differentiated from the Common Furniture Beetle by its more rounded appearance. It is always a little more reddish and is covered with short dense fair hair when freshly emerged. The lines of punctures along the wing-cases are not so deep. Householders often find the beetles, recognise their general similarity to pictures of Common Furniture Beetle, but find no flight holes. In such cases it is not difficult to track down the material which is infested, as often a few of the beetles are found hiding in the folds of the paper wrapping. It is a good plan to destroy the infested food, or include it in a chicken mash and then spray the inside of the cup-board and surroundings with an odourless oil-soluble insecticide. Do not use woodworm-killing fluids where food is stored.

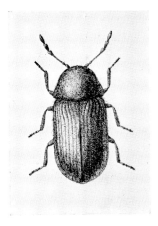

Fig. 69. The Biscuit Beetle, *Stegobium paniceum*, a near relative of the Common Furniture Beetle, but is not a wood-borer.

Plaster Beetles

A relatively large number of beetles have larvae which feed on fungal hyphae, so that where damp conditions favourable to the growth of fungi prevail, some species of these beetles may be attracted. They are sometimes present in enormous numbers in new houses or where new plastering has taken place. The damp plaster causes moulds or mildews to appear on paper, woodwork or textile fibres, and it is on these fungi that the adult beetle lays her eggs and on which the larvae subsequently feed. They are found also feeding on moulds growing on imperfectly-ventilated stored foodstuffs, such as jam or cheese, and in damp basements. The common name of Plaster Beetle is thus a misnomer.

Three species are very common, namely *Enicmus minutus*, *Lathridius nodifer* and *Cryptophagus acutangulus*. All these species (particularly the first-named) are found on woodwork and today are common timber-yard insects. Indeed, many timber-yard men confuse it with *Lyctus*, although *Enicmus minutus* is only a fraction of the size of *Lyctus*.

Plaster beetles do not damage woodwork.

Control (*indoors*): Dry out new plaster as soon as possible by keeping the rooms warm and well-ventilated. Wash newly plastered walls down with a fungicide, such as a solution of sodium pentachlorphenate in water, if it would appear that the moisture in the wall would dry out slowly.

With regard to timber (fresh sawn, 'wet' obeche seems very susceptible to *Enicmus minutus*) arrangements should be made to dry it out as quickly as possible.

Spider Beetles

A number of beetles in the family PTINIDAE, owing to their superficial resemblance to spiders, are known as Spider Beetles.

The species most commonly met with are:

Ptinus tectus   Australian Spider Beetle
Niptus hololeucus   Golden Spider Beetle
Ptinus fur   White-marked Spider Beetle

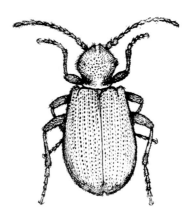

Fig. 70. The Australian Spider Beetle (*Ptinus tectus*), originally a native of Tasmania, now spread all over the world as a pest of stored foods.

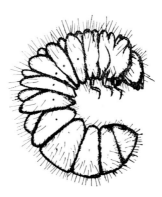

Fig. 71 The grub of the Australian Spider Beetle. Known to British soldiers during the First World War as the inhabitant and sharer of their Army biscuits.

They are all pests of stored products, but are often found in the older type of house, where the larvae subsist on bread-crumbs and similar farinaceous material lodged in crevices between the floorboards. They are however able to thrive on a wide variety of materials, including mouse droppings, and probably wool dust is also eaten by them. It is difficult to account for the occasional occurrence of swarms of the Golden Spider Beetle. The larvae of all the Spider Beetles are creamish in colour and roll themselves into a ball if disturbed.

Generally, it can be said that Spider Beetles do not damage wood, but occasionally the larvae do hollow out pupal chambers in wooden surfaces.

Control: Find the source of infestation and destroy, if possible, any food material containing the live stages. Spray regularly with oil-based insecticides all crevices in woodwork, etc., in the vicinity of the infestation.

Silverfish   *Lepisma saccharina*

The Silverfish insect is about half-an-inch long, cylindrical, rather torpedo-shaped and with three long bristles emerging from the tail-end. It is metallic grey in colour, perhaps more like lead than silver. It is very agile, seeking cover extremely rapidly when disturbed in its usual dark quarters. It is especially fond of warm damp places. It does not damage woodwork; indeed, it is seldom that any damaged material is observed which can be put down to Silverfish. It is closely related to an insect known as the Fire Brat, *Thermobia domestica*, often quite common in factories, bakehouses, etc., where steam heating is employed.

Control: Liberally dust with powder insecticide and use powder blower to distribute the powder in the crevices in brickwork and plaster used by the insects.

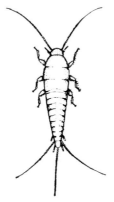

Fig. 72. Silverfish. Metallic grey in colour. Does no damage to woodwork and little damage at all.

H

Hide Beetles DERMESTIDAE

These beetles are notable pests of stored food products, skins, furs and materials of high protein content. A common species is *Dermestes lardarius*, known as the Larder Beetle or Bacon Beetle. The larva has a banded appearance and is beset with hairs of various lengths. The larva should never be handled, as some of the hairs break off in the skin and set up considerable irritation. Just before pupation takes place the larva burrows into any convenient compact material, and wood is frequently chosen. The diameter of the ingress hole and tunnel is about one-eighth of an inch, circular, and may be anything from about an inch to about a foot in length. Hide Beetle galleries in wood are however, only common in such places where the larvae are infesting dried skins or dried meat products such as in warehouses and tanyards. The banded cast larval skins are often present in the holes.

The adult beetles vary from seven to nine millimetres in length and are easily identified on account of the characteristic pattern of greyish-gold and black shown on the accompanying illustration.

Control: Thoroughly clean out premises where Larder Beetles occur and burn débris and infested material. Treat all exposed woodwork with a woodworm-killing fluid. Spray all materials susceptible to infestation regularly for a period—say, once a week for two months—with an oil-soluble insecticide.

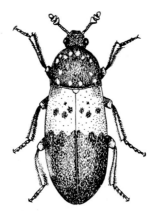

Fig. 73. The Larder Beetle, *Dermestes lardarius*, sometimes a wood-borer—when searching for a site for pupation.

Carpet Beetles

The larvae of Carpet Beetles (belonging to the beetle family DERMESTIDAE) are important pests of woollen material and furnishings in most parts of the world. Carpets are the usual furnishings attacked; that is why this group of insect pests is so named.

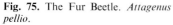

Fig. 74. Side view of a grub of *Attagenus* type of Carpet Beetle.

Fig. 75. The Fur Beetle. *Attagenus pellio.*

Fig. 76. The 'woolly bear' grub of *Anthrenus* type of Carpet Beetle.

Grubs of the *Attagenus* group are rather long and slender, greyish-brown or light-brown in colour, with a tuft of long hairs on the tail region. They are often found underneath the edges of carpets (especially fitted carpets) and the banded appearance of the cast or moulted skins is quite characteristic.

The adult beetle of *Attagenus pellio* is now usually referred to as the Fur Beetle. It is about three-sixteenths of an inch long, and black in colour except for three indistinct greyish marks on the prothorax and two distinct grey spots on the wing-cases (almost in the middle of the back).

On the other hand, the grubs of the *Anthrenus* group are much fatter in appearance, with long body hairs and several tufts of long hairs in the tail region. They are often known as 'woolly bears'. Their hairy appearance blends with the carpets or materials which they are attacking. In addition, they often hide in cracks in wooden floorings and clothes cupboards, etc. In airing cupboards they seem very resistant to dryness, and are able to withstand such dry conditions which clothes moth grubs could never tolerate. The adult beetles into which the grubs develop are rather globular, somewhat like the Ladybird Beetles (*Coccinella*). The various species, however, have different patterns of black, white and grey markings. The species illustrated (Fig. 77), is *Anthrenus verbasci*, usually called the Varied Carpet Beetle (Buffalo Bug or Buffalo Moth in North America).

The grubs of Carpet Beetles are very much more difficult to eradicate with complete success than the Clothes Moth, due partly to their habit of secreting themselves in tiny cracks in woodwork and partly to the scaly effect of their bristles, which seem to hold off powder insecticides and many sprays. But, if regular spraying of all woollen materials is carried out with a mothproofing fluid (specially formulated for moth control), eradication

Fig. 77. The Varied Carpet Beetle.

will be effected. Care should be taken, however, to spray over all—not to give a soaking in one place and nothing in another.

Furniture Mite and Book Louse

A few years ago Furniture Mite was a serious inconvenience to many house-holders. It is indeed strange to have to record that the Furniture Mite does not appear to be so common today as it was formerly. On the other hand, the Book Louse is today exceedingly common, and throughout the summer months innumerable requests are received for the identification and treatment in order to eradicate 'small creamy-white, quick-running insects about one-sixteenth of an inch long which run over the surface of the furniture and then hide in cracks and crevices'.

From this description it is easy to determine the cause of the annoyance as the Book Louse or *Psocid*. They have no connection, of course, with wood-boring insects such as the Common Furniture Beetle, although they are often found in association. They cannot be said to cause any damage to furnishings. It is believed that they feed only on minute fungi, which, invisible to the naked eye, occur on the surface of glues and pastes. It is a cause of considerable annoyance to the housewife to see those white insects running about on the furniture. In fact, the reaction of the average person on seeing them for the first time is that the insects are something of a much more destructive nature.

The life-cycle of the Book Louse is very simple. Eggs are laid by the females amongst dust in crevices in woodwork and after a short time the young Psocid hatches. It does not, like many insects, undergo a larval or grub stage but is in most respects very similar to the adult (except for size and sexual maturity). As it feeds, it changes its skin several times, and during the summer months takes only a few weeks to reach the adult stage.

To eradicate Psocids is fairly easy. They are usually indicative of warm, humid conditions prevailing, and all windows in a room where they occur should be kept open for several days, day and night for preference. Thereafter, treatment with the fluid normally used for woodworm treatment is sufficient to destroy them completely. The fluid should be applied by brush, special attention being paid to joints and cracks in the woodwork where the Book Lice will be seen to scurry out and succumb within a few seconds or so.

Tests carried out by the writer on Psocid destruction in antique furniture gave remarkable results which were both gratifying and extremely successful.

Conditions which favour the increase of Book Lice also favour the increase of the Furniture Mite, and perfectly dry furnishings (especially upholstered suites) brought into damp newly-built houses often bring about an attack by Furniture Mite. Furniture Mites are only just visible to

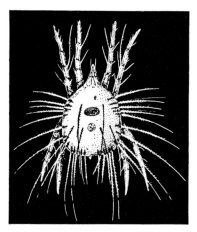

Fig. 78. Book Louse (*Psocid*). It feeds on minute fungi on damp furniture.

Fig. 79. Furniture Mite. Note eight legs and curiously branched bristles.

the naked eye and appear as minute white specks when seen on dark polished wood. They move very slowly, so that it is easy to differentiate between this pest (not an insect; it belongs to the spider group) and the much larger, quick-running Book Louse.

The Woodlouse   *Porcellio scaber* and *Armadillidium vulgare*

The Woodlouse is not an insect but a crustacean (in common with shrimps and crabs). It is characterized by its ten tile-like segments which articulate with each other and, on the underside, its seven pairs of walking legs. From the head arise a pair of large jointed antennae. It is rather variable in colour, from bluish-grey to greyish-brown. The species *Armadillidium vulgare* can roll itself in a complete ball when disturbed. Both species feed on vegetable débris of various sorts and frequent damp and dark places; in fact, very humid conditions are essential for its existence. This accounts for it being found in damp places in kitchens, near drains, and sometimes actually in the bath. It causes no damage to sound wood but is often found where wood is in the last stages of fungal decay. It is a nuisance but completely harmless in the home.

Fig. 80. The Woodlouse, an isopod crustacean. It feeds on vegetable debris, usually much decayed.

12

THE ANSWER TO THE PROBLEM OF WOODWORM IN BUILDINGS

Guaranteed Treatment by a Servicing Company—Woodworm Insurance—Pretreatment

Having now made a survey of the woodworm problem in the United Kingdom no one can be in any doubt concerning the overall importance of the Common Furniture Beetle and its grave significance in regard to timber used for constructional and joinery purposes in buildings. It has been said that man creates his own pests, and certainly in these islands, with regard to the little brown beetle, man has created the very conditions which allow this wood pest to flourish, to damage and finally to destroy.

What is the answer to this great problem? First, let us take the problem as it affects existing buildings.

Guaranteed Treatment by a Servicing Company

Having arrived thus far in the book, the reader will possess sufficient knowledge of the biology of wood-boring beetles to appreciate two important and fundamental facts regarding woodworm infestation by the Common Furniture Beetle, whether it concerns a piece of furniture or the structural timbers of a building.

First, flight holes merely show where the previous generations of beetles emerged. Secondly, flight holes do not indicate the total distribution of the present generation of larvae. These are boring away out of sight in the wood, and will probably continue to do so for another year or two before they, in turn, complete their metamorphosis and bite their way out of the timber to the outside world as adult beetles.

When it is remembered that the female lays from twenty to sixty eggs, sometimes over a wide area, it does not take much imagination to realize that the unseen distribution is usually far greater and probably more widespread than the original flight holes suggested. This fact alone immediately makes every woodworm attack, particularly where it is in structural timbers of a property, a two-fold problem. The first part of the problem concerns those timbers which show evidence of attack; i.e. the presence of flight holes. These timbers, and usually those in the vicinity of the attack, making up a separate building unit, i.e. the roof void, will be treated insecticidally.

Provided this treatment is carried out by a reputable firm specializing in this form of remedial treatment and provided also that the treatment is covered by an adequate twenty-year guarantee, then the property owner has solved the first part of the problem, and has nothing further to worry about, at least as far as the treated timbers are concerned. The virtual certainty of woodworm extermination, when carried out in this way has been commented on in a previous chapter.

We must now consider the second part of the problem, the remaining parts of the house not treated insecticidally but where larvae may be present although not visible as yet. To make this clear, let us imagine for a moment that the reader has just bought a house. It is about twenty years old, with an accessible roof void, four bedrooms, bathroom and the usual number of reception rooms on the ground floor.

As a result of the surveyor's inspection, a fairly heavy and widespread attack by Common Furniture Beetle was discovered in some of the rafters, joists and the boarding of the roof. One of the bedroom floors was also attacked, but not quite to the same extent. These timbers showed evidence of flight holes and were therefore known to be attacked, and on the recommendation of the surveyor, treatment was carried out to the roof timbers and the bedroom floor by a firm giving a twenty-year guarantee.

If we stop now for a moment and analyse the report, which stated that the attack in the roof timbers was fairly heavy and widespread, it must be obvious, first, that the attack is of more than one generation. Secondly, the infestation in the bedroom floor was probably a direct result of the attack spreading from the roof timbers. Remember, the beetles are free-flying insects, and the floor of the bedroom became infested due to the attack in the roof timbers. Once these two facts are put together we can then ask: how much further has the attack spread? What about the floors of the other rooms? This is the question, of course, that brings us to the second part of the problem; those timbers which appear to be free of infestation because there are no flight holes to be seen but which may in fact already be infested by the early larval stage. Even if they were not, these untreated timbers are susceptible to attack coming from elsewhere, and the possibility is that they may become infested at some time in the future. What can we do about these untreated and susceptible timbers? How can we safeguard them and also the pockets of the householder from possible further heavy financial outlay for remedial treatment in the future?

Woodworm Insurance

Fortunately, there is a solution to this second part of the problem, and what is probably more important, an answer both simple and inexpensive. It is to insure against it. To be able to insure one's house against woodworm damage is a relatively new idea in this country, but in Scandinavian countries it has been possible to do so for a number of years; in fact, in Denmark, woodworm insurance has been in force for more than thirty years. In Denmark, however, the House Longhorn Beetle is the hazard. In Britain this form of insurance is effected through an Insurance Broker or merely by completing a proposal form. The Insurance Company will then arrange to carry out a free preliminary survey of the timbers of the property. If no evidence of infestation is revealed then the policy is written without further delay. On the other hand, if there is evidence of an attack, then the infested areas must be treated before the insurance can be granted. However, due allowance is made in the annual premium for any treatment which has had to be carried out prior to the granting of insurance.

One might ask, what are the benefits to be derived from this form of insurance? First, the owner of the property is relieved from worry and future financial outlay, apart from the fact that the capital invested in the property is protected at reasonable premium rates. Many householders would also regard the free preliminary inspection as a benefit, and the fact that the property is insured against attack must surely be an added selling point, should the owner decide to sell his property. A woodworm insurance policy is also assignable by notifying the Insurance Company of the change of ownership.

Pretreatment

Finally, what of the future? Should anything be done about the 250,000 dwellings built every year with a standard or so of timber in each? There is only one reasonable answer to this question, of course; timber should be so treated as to render it immune from woodworm attack before the timber is fixed into position. The period of effectiveness should be for as long as scientific research can make possible, and there is reason to hope that even now some of the products and processes available for *Pretreatment*, as it is called, would preserve timber indoors in buildings for the length of life of the building itself.

Although a few pre-treatment processes suitable for building timbers have been available for some years, they have only been used on a relatively small scale. It is hoped that this situation will improve. There are two main types of process: vacuum/pressure impregnation and dipping. More information on vacuum/pressure impregnation is given in two other books in the Rentokil Library Series, 'Wood Preservation—A Guide to the Meaning of Terms' and 'Termites—A World Problem'. Some timber is being imported into the British Isles having been dipped with borax solution when the timber was green, i.e. very wet. The borax solution is able to diffuse into the wet wood.

For a pre-treatment process to enjoy univeral use it seems self-evident that (1) it should be simple in operation, (2) water should be the solvent used for reasons of economy and (3) the water-soluble salts should be very soluble but should become fixed to the cellulose of the timber. There are several methods which satisfy these requirements, for example the Rentex process, a dipping treatment, and the Celcure process, a vacuum/pressure treatment.

BIBLIOGRAPHY

Below is given a selection of books, leaflets and papers in which the reader will find various aspects of the woodworm problem dealt with in greater detail than is possible in the present book.

FOREST PRODUCTS RESEARCH LABORATORY LEAFLETS (H.M.S.O., London).
Beetles injurious to timber and furniture. (No. 19, 1940).
Defects caused by Ambrosia (Pinhole Borer) Beetles (No. 50, 1956).
The Kiln Sterilization of *Lyctus*—infested timber (No. 13, 1957).
The House Longhorn Beetle (No. 14, 1958).
The Common Furniture Beetle (No. 8, 1959).
The Death-Watch Beetle (No. 4, 1959).
Lyctus Powder-Post Beetles (No. 3, 1959).
Prevention of *Lyctus* attack in sawn Hardwoods by use of 'Contact' Insecticides (No. 43, 1960).

BLETCHLY, J. D. The Influence of Decay in Timber on susceptibility to attack by the Common Furniture Beetle, *Anobium punctatum* de Geer. *Ann. appl. Biol.*, **40** (1), 218–21 (1953).

BLETCHLY, J. D. The Biological Work of the Forest Products Research Laboratory, Princes Risborough III. The Works of the Entomology Section, with particular reference to the Common Furniture Beetle, *Anobium punctatum* de Geer. *Proc. Linn. Soc. Lond.*, **168**, 111–15 (1957).

BLETCHLY, J. D. The Influence of Soft Rot on the Susceptibility of Beech to attack by Common Furniture Beetle (*Anobium punctatum*). *British Wood Preserving Association Convention Record*, p. 5 (1959).

BRITTON, E. B. Domestic Woodboring Beetles. *Brit. Mus. (N.H.)* Economic Series No. 11a (1961).

DUFFY, E. A. J. A Monograph of the Immature Stages of the British and Imported Timber Beetles (CERAMBYCIDAE). *Brit. Mus. (N.H.) Lond.*, (1953).

FISHER, R. C. Studies of the Biology of the Death-Watch Beetle, *Xestobium rufovillosum* de Geer IV. The Effect of Type and Extent of Fungal Decay in Timber upon the Rate of Development of the Insect. *Ann. appl. Biol.*, **28**, 244, (1941).

GARDINER, P. The Morphology and Biology of *Ernobius mollis* L. (COLEOPTERA ANOBIIDAE). *Trans. R. ent. Soc. Lond.*, **104** (1), 1–24, (1953).

HARRIS, W. V. *Termites, their recognition and control.* Longmans Green, London, (2nd Ed. 1972).

HICKIN, N. E. *Woodworm, its Biology and Extermination.* Simpkin Marshall, London, (1954).

HICKIN, N. E. *The Insect Factor in Wood Decay.* Hutchinson, London, (2nd Ed. 1968).

HICKIN, N. E. *The Conservation of Building Timbers*. Hutchinson, London (1967).

HICKIN, N. E. *Wood Preservation—A guide to the meaning of terms*. Hutchinson, London (1971).

HICKIN, N. E. *Termites—A World Problem*. Hutchinson, London (1971).

KELSEY, J. M. Symbiosis and *Anobium punctatum* de Geer. *Proc. R. ent. Soc. Lond.* A. **33,** 21–4, (1958).

LINSCOTT, D. The Susceptibility of Imported and Newly-seasoned and Home-grown Hardwoods. *J. Inst. Wood Sci.* **28,** 36–9, (1971).

MUNRO, J. W. The Larvae of the Furniture Beetles. *Proc. R. phys. Soc. Edinb.*, **19,** 220–36; (1915).

PARKIN, E. A. The Larvae of some Woodboring ANOBIIDAE. *Bull. ent. Res.*, **24,** 33–68, (1933).

PARKIN, E. A. The Digestive Enzymes of some Woodboring Beetle Larvae. *J. exp. Biol.*, **17** (4), 364–77, (1940).

SNYDER, T. E. *Our Enemy the Termite*. New York, (1948).

SPILLER, D. Note on the Susceptibility of *Pinus radiata* D. Timber to attack by the Common House-borer *Anobium punctatum* de Geer. *N.Z. Sci. Tech. B.*, **32** (5), 38–9, (1951).

WHITE, M. G. The Status of the House Longhorn Beetle. *Timber Technology*, **67,** 406–11, (1959).

INDEX